T0413210

Theories on Law and Ageing

Israel Doron
Editor

Theories on Law and Ageing

The Jurisprudence of Elder Law

Editor
Dr. Israel Doron
University of Haifa
Department of Gerontology
Mount Carmel
31905 Haifa
Israel
idoron@univ.haifa.ac.il

ISBN 978-3-540-78953-6 e-ISBN 978-3-540-78954-3

DOI: 10.1007/978-3-540-78954-3

Library of Congress Control Number: 2008925884

Cover Design: WMX Design GmbH, Heidelberg, Germany

Printed on acid-free paper

5 4 3 2 1

springer.com

Preface

This book is about trying to answer questions. These questions were well introduced by Prof. Margaret Hall in the opening of her chapter in this book:

> "The fundamental idea of 'law and aging' as a discrete category of legal principle and theory is controversial: how and why are 'older adults' or 'seniors' or 'elders' (the very terminology is controversial and fraught with difficulties) a discrete and distinct group for whom 'special' legal thought and treatment is justified? For some, a category of law and aging is inherently paternalistic, suggesting that older persons are, like children, especially in need of the protection of the law.
>
> In this sense, the argument continues, the category itself internalizes ageist presumptions about older adults and is therefore inherently flawed and even harmful. If certain older adults are, because of physical or mental infirmities, genuinely in need of an enhanced level of legal protection, this entitlement should be conceptualized in terms of their disability; older adults are not a distinct group but an arbitrarily delineated demographic category which contains within it any number of groups that are legitimately distinct for the purposes of legal theory (the disabled; women; persons of colour; Aboriginal persons; rich and poor; etc.) Indeed, the artificial category of "older adults" may be seen as obfuscating, submerging these more meaningful distinctions.
>
> The essential question underlying any theory of law and aging is, therefore, this question of conceptual distinction. What special feature and characteristics of "older adults" justify and even require a particular theoretical approach? Is it possible to formulate legitimate generalizations about a group identified as "older adults," while avoiding the harmful stereotypes of ageism? And what if anything is gained by that approach?"

While elder law and the research and study of the older population has grown substantially in the last three decades, only a few attempts have been made to conceptualize or theorize this field of law in a systematic manner. Similar to the field of gerontology, (e.g. Bengston et al. 2005), there are various good reasons for failing to theorize: existing jurisprudence already provides ample theoretical basis; abstract theory is mostly irrelevant to lawyers and attorneys who represent older clients on daily basis; one does not need theory to study and conduct legal research in this field; and finally, theory could actually become, at least sometimes, a limitation to our understanding and places unnecessary borders to the development of this field.

However, in this book, we beg to differ. The authors in this book share the belief that developing, discussing and providing a theoretical frame work to the field of elder law, has become necessary. When we use the word "theory" we use it as a broad term: theory in this book means any construction of explicit explanation that accounts for the reality (in its empirical sense). Every one who works, studies or is involved in the legal field of law and ageing, encounters some kind of a legal reality: court rulings, parliamentary legislation, administrative rulings, family agreements, and much more. We need theory in order to systematize what we know, to conceptualize the "how" and the "why" behind the "what" that we are doing (Bengtson et al. 1999).

Without a theoretical framework, we cannot justify the existence of elder law as a distinct field within law. Theory provides us with the necessary intellectual tools to evaluate and put to a test what we are doing. In the words of Bengtson et al. (1999, p. 5), theory provides "a set of lenses through which we can view and make sense of what we observe in research." It is through this lens that we can eventually build knowledge and understanding in a systematic and cumulative way, so that our empirical efforts will lead to integration with what is already known as well as a guide to what is yet to be learned (Bengtson et al. 1996).

In an attempt to stand up to the challenge of presenting different theoretical frameworks in the field of law and ageing, each chapter in this book will try to present a different theoretical perspective, which contains the answers to the questions posed above. The first chapter, written by the American attorney Allan Bogutz, provides the historical perspective for the theoretical discussion. As an active elder law attorney, and as one of the founders of NAELA (the National Academy of Elder Law Attorneys), advocate Bogutz has experienced in person the struggles that surrounded the need to define and justify the creation of a distinct legal specialty. His historical overview and personal insights provide an excellent basis for the understanding of the rest of the book.

The second chapter in this book, written again by one of the leading American scholars who have followed the development of elder law throughout time, is Prof. Lawrence Frolik. Prof. Frolik presents one of the most well known and important approaches to elder law: elder law as later-life planning. As he describes, "[i]n response to client needs, elder law has expanded and is gradually redefining itself into "later life planning." While some still identify elder law with helping clients pay for long-term care, specifically in the United States by qualifying for Medicaid, the reality is that the practice of elder law is a rich mosaic of legal planning that is continually evolving to better meet client legal, financial and social needs and concerns."

"Later life planning" is a term that hides a richness of legal tools, mechanisms and instruments. Starting from property management and long term health care planning, this concept has broadened to include old-age housing issues, retirement planning, health-care decision making, and advanced legal planning for mental incapacity. Prof. Frolik describes in his chapter these different issues, and the ways law can become an efficient legal planning tool for adults and older persons.

In the next chapter, Prof. Marshall Kapp argues that the proper theoretical framework in which elder law should be placed is Therapeutic Jurisprudence (TJ). Prof. Kapp has written and studied the different regulatory aspects of law and ageing in an extensive manner. His prolific writing in this field has led him to adopt a relatively novel approach to elder law, one that focuses on prevention and on therapy. As described by Prof. Kapp, "TJ [Therapeutic Jurisprudence] suggests the law and legal practice have inevitable consequences for the well-being (psychological, physical, financial, and other) of involved people. It is an interdisciplinary, interprofessional field of legal scholarship – a particular analytical lens – with a pragmatic, realist law reform agenda. TJ seeks to identify the therapeutic and antitherapeutic (and therefore counterproductive) effects of law and involvement in the legal process, and to shape the law and legal practice in ways that diminish the antitherapeutic consequences and maximize the therapeutic potential for actual, identifiable (as opposed to abstract) older individuals."

Prof. Kapp provides in his chapter ample examples on how TJ is applied in real life in different spheres of elder law, and how adopting the TJ "analytical lens" can provide different insights on the appropriate regulatory schemes in the field of law and ageing. His conclusion is that "law is not an end in itself; rather, it is a necessary instrument to move societies toward desired practical goals. TJ is a valuable tool for telling us how well or poorly the law is doing in moving us toward the goal of improving the quality of life for older persons."

From therapy and prevention, the theory moves to the field of feminism and the feminist perspective on law and ageing. As analyzed by Prof. Kim Dayton, yet another central pillar of the academic field of elder law in the US, a feminist analysis of law and ageing is crucial to this field. This is so not only because of demographics: women consist the majority of the older population, especially that of the older-old and the poor. It is also because feminist legal theory is actually rich and diverse as to include a critical perspective on the ways law constructs old age – in general, and elderly women – specifically.

The next chapter, by Dr. Israel (issi) Doron, tries to escape the "monist" approaches: the key for conceptualizing elder law is not found in one dimension or a single perspective. Rather, a multi-dimensional model is needed in order to encompass this dynamic field. Known for his international and comparative perspectives on law and ageing, Dr. Doron provides a model which consists of five different "legal dimensions," which are inter-related and inter-connected.

The core of Dr. Doron's model is based on the fundamental constitutional and legal principles of the existing legal system by means of which the rights of the old can be defended and grounded in law, even though they contain no specifically age-related provisions; The protective dimension aims at protecting the elderly population against abuse and injury; The dimension of family support strengthens informal social reinforcement networks; The planning and preventive dimension attempts to put into practice the principle of the individual freedom of the aged and to enable them to realize their desires and aspirations even when they are no longer competent or in control of their faculties; And, finally, the dimension of empowerment includes techniques of education, explanation and representation without which old people are incapable of exercising their legal rights. In the chapter, different examples are given to the applications of the model in real elder law cases.

The sixth chapter is by Prof. Richard Kaplan. Prof. Kaplan has been one of the leaders in adopting an already well known and well established approach to law, the law and economics approach, and implementing it to the field of law and ageing.

As presented by him, the common understanding of this approach is that "[e]conomics provide[s] a scientific theory to predict the effects of legal sanctions on behavior" and that "economics provides a behavioral theory to predict how people respond to changes in laws." Moreover, economics is able to accomplish this important function, its proponents assert, because it "has mathematically precise theories (price theory and game theory) and empirically sound methods (statistics and econometrics) of analyzing the effects of prices on behavior."

Prof. Kaplan takes this well known theoretical approach and applies it to the field of law and ageing. In different legal examples, which are described in detail in his chapter, he shows how economic analysis of legal policies regarding the older population is crucial to the understanding and evaluation of any legal regime in this field. Yet, this application is not done without a critical observation of the limits and hazards that one has to take into account once adopting this approach to the field of law and ageing.

The next chapter by Prof. Margaret Hall. Prof. Hall, one of the leading Canadian scholars in the field, traces the key concept within law and ageing as "vulnerability," or the resistance to this concept. In her words,

> "[r]esistance to the idea of vulnerability as key to a conceptually coherent category of "law and aging" is strong, and rooted in the idea that vulnerability = weakness and resistance to the presumption that age = loss of capacity. The fear is that legal theory focusing on personal vulnerability increases social vulnerability, the more significant source of harm, to the extent that it reinforces ageist presumptions of weakness and incapacity. Legal protection for the truly incapable, of whatever age, exists; and beyond that, older adults should be treated in law and otherwise like any other adult persons."

Prof. Hall finds her theoretical "solution" to the "vulnerability" challenge in an original application of well known legal concepts. "The key is to rethink our ideas of vulnerability in this context, drawing on the conceptual framework of equitable fraud: the venerable doctrines of undue influence, and unconscionability. Equity provides a coherent and sophisticated theoretical model for understanding vulnerability as both situational and relational, as opposed to the capacity/autonomy duality. The doctrines are related, but distinct. The inequity of the unconscionable transaction lies in one person's exploitation of the vulnerability of another. The inequity of undue influence lies in the effect of one person's "undue" influence of the ability of another to give free and independent consent. In neither scenario does the "weaker" part lack capacity; vulnerability arises through the power dynamics of the relationship together with factors of dependence and inequality." Prof. Hall provides ample examples and evidence of how her theory actually works in real life and with real older people.

Elder law has been challenged by those who have argued that it is nothing more than another branch of disability law. Prof. Doug Surtees, a Canadian expert in the field of disability law, does not adopt this point of view. However, through his own expertise in the field of disability law, Prof. Surtees describes the historical development of the disability rights movement, and the different mile-stones along its way. Taking up this rich experience, Prof. Surtees argues that lessons learned from the disability movement are relevant to the field of elder law. Most importantly, the concept of "universalism," which has been developed in the legal realm of disability rights, is of much relevance to the conceptualization of elder law.

As asserted by Prof. Surtees, "Universalism as a model to understand elder law carries with it the hope that all of us can be united in designing programs and policy which include us all, wherever we currently find ourselves on time's continuum. Age is a circumstance which, when combined with some life situations, can impact upon a person's vulnerability. Age should not be ignored. The civil rights model, however, risks using age to divide us. It uses age to determine if a certain person is an 'elder' or not, in the same way as it asks if a person is 'disabled' or not. It is divisive. When we are divided some are marginalized. A continuum of age should be used for inclusion, not to divide us. Universalism holds this promise."

Before closing, Prof. Winsor Schmidt presents another, yet different, theoretical analysis to the field of law and ageing. Prof. Schmidt has for many years studied the legal instrument of adult guardianship, and his research and empirical findings serve as the basis for significant law reforms in this field. His mental-health perspectives provide another unique and original lens to understand the uniqueness of elder law. As described by Prof. Schmidt, from a critical perspective, law and aging can be seen as a "potential mechanism of socially controlling the social deviance of aging through the therapeutic state and therapeutic jurisprudence."

However, he proposes to adopt an alternative mental health theory approach to law and aging, which is normatively based on the premise of free will and responsibility and a legal mental health system that abolishes involuntary civil commitment and has sanism as a principal challenge. From this original theoretical perspective, it is clear that the challenge elder law faces in the future is to transform itself from a social control mechanism of a therapeutic state and a therapeutic jurisprudence into an emancipating, empowering, integrating assertion in an aging society.

The book is closed by a concluding chapter written by one of the leading scholars in the field of elder law, Prof. Rebecca Morgan. Prof. Morgan, provides in the final chapter a personal vision on the future conceptual development of the field of law and ageing. In her words: "aging is a great universal – everyone does it – every day, but everyone does it differently. No one solution, no one theory, no one-size-fits-all approach to Elder Law will serve. Instead, take the best of all and craft the best solution – or solutions – for the client – or clients. Remember that aging is universal – since everyone does it every day, each person has a vested interest in the future of Elder Law."

This challenge which is set in her chapter, establishes a platform through which we can relate to, once we will look back in retrospect, and try to see what, if any, of the theoretical approaches presented in this book, has actually proven to be valid, or successful, in advancing the knowledge and understanding of the field of law and ageing.

It is our hope that this book will serve its goal: it will serve as a basis for further discussions and developments in conceptualizing what is done in the field of law and ageing. We hope that the richness of ideas, of legal perspectives and of original point of view that are presented in this book, will force any one who is involved in legal issues of older persons, to try and place his or her action within a broader theoretical framework. We hope that this book is only a beginning: a beginning of a new legal road, which will look at law and ageing as a rich, full and exciting field of law that holds a coherent – yet diverse – conceptual basis for future growth.

Special thanks are given to Mrs. Brigitte Reschke, Senior Editor of Law, and the Springer Publication for all their assistance and support for publishing this book.

February 2008 Israel Doron

References

Bengtson VL, Parrott TM, Burgess EO (1996) Progress and pitfalls in gerontological theorizing. Gerontologist 36:768–772

Bengtson VL, Putney NM, Johnson ML (2005) The problem of theory in gerontology today. In: Johnson ML (ed) The Cambridge handbook of age and ageing. Cambridge University Press, Cambridge, pp 3–20

Bengtson VL, Rice CJ, Johnson ML (1999) Are theories of aging important? Models and explanations in gerontology at the turn of the century. In: Bengtson VL, Schaie KW (eds) Handbook of theories of ageing. Springer, Berlin Heidelberg New York, pp 3–20

Contents

List of Contributors

Bogutz, Allan D. – Advocate and a private attorney, practices elder law with the firm of Bogutz & Gordon, PC, in Tucson, Arizona. He is one of the founders of NAELA and a past president, a Fellow and a Certified Elder Law Attorney. He has been recognized as one of the Best Lawyers in America for 2007 and 2008 and is a Distinguished Scholar of the Canadian Centre for Elder Law Studies. He is a former commissioner of the ABA Commission on Law and Aging and a Fellow of the American College of Trust and Estate Counsel.

3503 N. Campbell Avenue Suite 101, Tucson, AZ 85719, USA
aginglawyer@gmail.com

Dayton, A. Kimberley, – Professor of Law at the William Mitchel College of Law. She is a co-author of *Advising the Elderly Client* (Thomson-West), a four-volume, 39-chapter treatise on elder law and related topics (with Guare, Mezullo, and Wood), and *Elder Law: Readings, Cases, and Materials* (Lexis-Nexis 3d ed. 2007) and its companion statutory supplement (with Gallanis and Wood). In 1994, Professor Dayton helped to establish the elder law clinic at the University of Kansas, one of the nation's first law school clinics dedicated exclusively to serving the needs of elderly clients. She founded the Kansas Elder Law Network, a web-based compendium of resources on elder law, in 1995. In 2003, KELN was renamed the National Elder Law Network, www.neln.org.

William Mitchell College of Law, 875 Summit Avenue, St. Paul, MN 55105-3076, USA
kim.dayton@wmitchell.edu

Doron, Israel – Professor of Law, senior lecturer at the Department of Gerontology, Haifa University, Israel and the founder of "The Law in the Service of the Elderly" association in Israel. A Distinguished Scholar of the Canadian Centre for Elder Law Studies and the recipient of the Zusman Award for excellence in promoting the rights of the elderly in Israel. He authored several books in the field of elder law, including *Age, Law and Justice* (in Hebrew).

Department of Gerontology, Faculty of Welfare and Health Sciences, Haifa University,
Mount Carmel, Haifa, Israel 31905
idoron@univ.haifa.ac.il

Frolik, Lawrence, A. – Professor of Law at the University of Pittsburgh School of Law. A frequent lecturer on legal issues of aging, and a prolific author, his books include, *The Law of Later-Life Health Care and Decision Making* (ABA Press) *Advising the Elderly or Disabled Client,* (2nd ed. with Brown) (Warren, Gorham & Lamont), *Elder Law in a Nutshell,* (4th ed. with Kaplan) (West) and the casebooks, *Elder Law: Cases and Materials,* (4th ed. with Barnes) (LexisNexis) and *The Law of Employee Pension and Welfare Benefits* (2nd ed. with Moore) (LexisNexis).

University of Pittsburgh School of Law, 3900 Forbes Avenue, Pittsburgh, PA 15260, USA
frolik@pitt.edu

Hall, Margaret, I. – Assistant Professor in the Faculty of Law at the University of British Columbia, and an Associate Researcher with the Centre for Research on Personhood in Dementia (located at the University of British Columbia). Margaret Hall has researched and published extensively in the area of law and aging, where her focus interests include material exploitation, capacity, housing regulation and inter-generational family relationships. Margaret also works in the area of tort law, focusing on public authority liability and family law generally, and is particularly interested in the position and protection of vulnerable individuals.

1822 East Mall, University of British Columbia Vancouver, BC, Canada V6T 1Z1
hall@law.ubc.ca

Kaplan, Richard, L. – Peter and Sarah Pedersen Professor of Law, University of Illinois, Champaign, Illinois, USA, and specializing in the areas of federal income taxation and policy and elder law. In addition to his numerous books and articles involving taxation and tax policy, he is the co-author of *Elder Law in a Nutshell,* published by West Publishing Co. (4th ed. 2006 with Frolik), as well as articles on various elder law topics, including Social Security, Medicare, long-term care financing, and retirement funding. He has served as the faculty advisor to *The Elder Law Journal*, the oldest scholarly publication devoted to this subject, since that publication was created in 1992. He was a delegate to the National Summit on Retirement Savings and is a member of the National Academy of Social Insurance.

College of Law, University of Illinois at Urbana-Champaign, 133 Law Building MC-594,
504 East Pennsylvania Avenue, Champaign, IL 61820-6996, USA
rkaplan@law.uiuc.edu

Kapp, Marshall, B. – Garwin Distinguished Professor of Law and Medicine, Southern Illinois University School of Law. Professor Kapp is the author or coauthor of a substantial number of published articles, book chapters, and reviews. He is the founding editor of the *Ethics, Law, and Aging Review* formerly published by Springer Publishing Company and founding editor of Springer's Book Series on Ethics, Law and Aging. Additionally, he is the present editor of the *Journal of Legal Medicine,* the official scholarly publication of the American College of Legal Medicine. In 2003, he received the Donald Kent Award of the Gerontological Society of America for exemplifying "the highest standards for professional leadership in gerontology through teaching, service, and interpretation of gerontology to the larger society."

School of Law – Center for Health Law and Policy, Lesar Law Building – Mail code 6804,
Southern Illinois University Carbondale, 1150 Douglas Drive, Carbondale, IL 62901, USA
kapp@siu.edu

Morgan, Rebecca C. – The Boston Asset Management Faculty Chair in Elder Law and Director of the Center for Excellence in Elder Law at Stetson University College of Law. Professor Morgan teaches a variety of elder law and skills courses, and oversees the Elder Law concentration program for JD students and is the director of Stetson's on-line LL.M. in Elder Law. She is a successor co-author of Matthew Bender's *Tax, Estate, and Financial Planning for the Elderly* and its companion forms book, and a co-author of *Representing the Elderly in Florida*. She is a member of the elder law editorial board for Matthew Bender.

Center for Elder Law, Stetson College of Law, 1401 61st St. South, Gulfport, FL 33707, USA
morgan@law.stetson.edu

Schmidt, Winsor C. – Professor of Health Policy and Administration and Chair of the Department of Health Policy and Administration at Washington State University. His research interests center on issues related to guardianship and the elderly, health and mental health law and policy, and medical malpractice. His publications include the books, *Public Guardianship and the Elderly* (Ballinger Publishing Company, 1981) and *Guardianship: Court of Last Resort for the Elderly and Disabled* (Carolina Academic Press, 1995), and numerous articles on health and mental health law. He is currently a member of the Washington Certified Professional Guardian Board, and the Board of Directors of the National Committee for the Prevention of Elder Abuse.

Department of Health Policy and Administration, Washington State University, P.O. Box 1495,
Spokane, WA 99210-1495, USA
schmidtw@wsu.edu

Surtees, D. – Assistant Professor at the College of Law, University of Saskatchewan (Canada). He teaches several courses including Elder Law and Disability Law. Doug lives in Saskatoon with his wife, Cathy and their daughters, Lindsay and ShayAnne.

College of Law, University of Saskatchewan, 15 Campus Drive,
Saskatoon, SK, Canada S7N 5A6
doug.surtees@usask.ca

Chapter 1
Elder Law: A Personal Perspective

A.D. Bogutz

My first "elder law" client was a 82-year-old woman living in a small, downtown, rent-subsidized one-bedroom apartment. The landlord had called the police to say that the woman whom we will call Annie had not been out of her room in more than a week. The police agreed to do a "welfare check" and arrived the next afternoon, knocked on the door and heard a faint "Come in." They entered the apartment and found Annie in bed in an apartment that was in disarray to say the least. Annie was weak but was able to get out of bed and sit in her recliner chair to talk with the police. She was quite alert. The police asked if she was all right and she replied quite frankly and straightforwardly telling them that she thought she was dying. The police tried to persuade her to go to the hospital but Annie was adamant that she wanted to remain in her apartment. She said she knew she had cancer of the lung, had had many series of treatments and now knew she was going to die; she simply wanted to die in her own bed.

The police called the Adult Protective Services office and an intake worker met with Annie and also tried to persuade her that she should be hospitalized. Annie remained adamant that she preferred to remain at home. She had all she needed, food delivered, a visiting nurse to administer pain medications and her rent was paid up. Why should she move? Adult Protective Services called the Public Guardian who went to court to establish a guardianship so that Annie could be moved from her home to the hospital against her will; the petition to the court alleged that Annie lacked capacity to make a responsible decision as to her care. This was when I was appointed by the court as the lawyer to represent Annie. Annie had no family; she had never married and had outlived her siblings.

When I met with Annie, she knew all about her illness, her impending death and I found her quite clear about her wishes. There was no one for her to turn to but she had daily home health visits and did indeed have home delivered meals. There was trash in the apartment but no smell of garbage or human waste. Annie said she could still get to the bathroom most of the time and slept on a rubber sheet "just in case." I was able to explain to her that the Public Guardian had filed a petition to become her guardian and hospitalize her, that there was a hearing in two days because the Public Guardian had said it was an emergency and that the petition said she did not have the ability to make her own decisions. I told her my job, as the court-appointed attorney, was to represent her in the hearing and to make sure that

I. Doron (ed.) *Theories on Law and Ageing: The Jurisprudence of Elder Law*,

her position on the petition was heard by the court. Did she understand? "Yes," she said, "They think I am crazy because I want to die at home. Is there a law that says I have to die in a hospital?"

I told her that there was no such law and that I was going to ask her some questions to check on what she knew about her condition and things in general such as time, place and current events. Annie was able to answer all of the questions appropriately and I learned that she had been an office manager for a large firm before she had moved to Tucson to retire more than 20 years earlier. She had a reasonable income and a fair amount of money in savings. She had a will that left her assets to a distant nephew and to charity. She was aware of her health issues and she was clear that she did not have long to live. With her money and with her care by a nurse, she felt that she would be able to be supported in the end stages of her illness in her apartment. I told Annie I would, as her lawyer, take the case to court.

In short, I was able to have a physician examine Annie and report to the court that Annie was fully able to make responsible decisions about her person and her care and that she should be left to live or die as she pleased. The court dismissed the Public Guardian's petition despite the Public Guardian's intake worker's testimony that Annie should be hospitalized. Annie died about 4 weeks later, in her own bed.

This elder law practice really is something new. Elder law became an area of law practice in the US in the mid-1980s. The reason there was no elder law sooner is that there were far fewer elders. In addition, there have historically been far fewer age-specific laws, pension programs and benefits.

Furthermore, trusts and estates law has always been an important area of law practice. People have always died at some point and, for almost all of history, there have been tax issues associated with wealth transfer on death. But death occurred much earlier for most people. In 1900, the average life expectancy in the US was 46 years. Today it is exceeds 80 years.

Longer life for a larger proportion of the population has been only one of the issues that led to the development of an area of law dedicated to the legal issues of aging. Attitudes about old age have taken some serious adjustment. In one of my cases, I had a physician write to the court that his patient should have a guardian appointed "because she is over 70 years old." Indeed, until the mid-1980s, "advanced age" was one of the criteria for the creation of a guardianship under Arizona law. Old age is not a disability!

People are living longer but many are living longer in a state of some degree of physical or mental disability. These people require care and incur the expense of such care. Medicine is able to keep people alive much longer than earlier, frequently alleviating many of the symptoms of illness and curing many illnesses that have historically been fatal. Our grandparents got sick and the physician comforted them; their illnesses progressed and they died, often at home. Today, treatment is aggressive, expensive and usually hospital-based.

I have had many clients over the last 30 years who have wanted to assure that they did not "hang on" or be a "burden" to their families. One of the services we have been able to provide to clients is the assurance that a living will along with a properly-appointed agent/surrogate/proxy will be able to protect them from unwanted care they may be able to personally refuse. There have been a number of times when I have been required to argue with physicians about the enforceability of such an advance directive and have several times had to resort to terminating the physician's services to assure compliance with the clients' wishes. I have had some clients who were former prisoners of war who wanted nutrition withheld but wanted hydration provided based on what they had seen in the prison camps. Other clients who have had a fear of starving to death were insistent that food be provided but that water could be withheld. Many clients, like Annie, have been adamant that they did not wish to be removed from their homes if they were dying.

Older persons often move at the time of retirement, moving towards leisure communities in salutary climates. When they move, they leave behind lifelong friends, community connections, religious institutions and other local resources that would be able to serve them in older life. They also leave behind natural family support systems. And if the elders are not relocating, often their children are moving to find work or to be near better schools or recreational opportunities. In either case, the older person who becomes disabled often finds himself or herself isolated in a community with which there are no natural ties or in an aging community where many cohorts face the same issues and create stresses on caregiving services.

In these situations, first as Public Fiduciary for Pima County, Arizona, and, for the last 27 years in private practice, I have been able to serve as fiduciary for such persons, acting personally as guardian, conservator (guardian of the estate), trustee, agent and/or administrator/executor for persons with no one. This has perhaps been the most gratifying type of work I have done and has led to long-term and very personal relationships with many clients and with some of their families. As the practice has grown, serving as fiduciary for these clients has expanded to the point at which my firm is now multidisciplinary, with lawyers, care managers who are social workers or registered nurses, financial managers who assist with financial management, bill paying, tax compliance and providing investment oversight for the clients' resources.

In addition to serving as personal guardian or estate manager for clients with no one able to serve, we have also been assisting in families where there are conflicts as to the type of care to be provided; we have been able to act as a disinterested professional in families with dysfunctions. An example is the client who had signed an advance directive that directed that no heroic measures be used to extend her life beyond its natural end; the client's son agreed to withdrawing food and nutrition from his mother while her daughter went to court to become guardian to overrule their mother's expressed wishes. In that case, the court appointed me to act as a professional fiduciary (on the nomination of the son) and I directed that the mother's wishes be honored. And, even for those who remain near family connections, two-income families now mean that there is no child or child's spouse at home during the day to provide necessary assistance.

So longer life, better medical care, geographic dispersal, disconnection from family care options and the disappearance of the family caregiver all lead to a greater dependence on governmental services for care and greater dependence on strangers for financial management and medical and residential decisionmaking in later life.

From my own personal perspective and professional experience, it is in this context that Elder Law evolved. Historically, older persons received care from their families, had long-term relationships with their physicians and had extended families and community connections that allowed an informal system of gradual assumption of financial management and personal decisionmaking on behalf of an elder who became unable to handle such matters for himself. Undoubtedly, such informal, familial or community caregiving continues in many places in the world but the urbanization of society and the busyness of life prohibits the commitment that such a caregiving, supporting undertaking requires. Internationally, the issues are the same as elders live longer and their natural support systems become unable to devote the time necessary to assist them.

Into this void, inevitably, there have flowed those who would take advantage of vulnerable older adults. Whether salespeople selling nonexistent or unneeded products or services, salespeople selling an illusion of needed insurance, charities overreaching to urge donations, lenders offering deals that are too good to be true or caregivers using undue influence to take from the elder either during life or by will, the vulnerable adult is indeed in need of some protection and advocacy.

I have been involved in a number of cases where recovery of assets was the issue. Too frequently, these cases have involved vulnerable older persons who were taken advantage of by either family or strangers. In one memorable case, a gentleman befriended an older widow and persuaded her that the local law prohibited women from owning certain types of property, including real estate. Trusting him, the widow signed over all of her real estate, including her home, to the man so that he could protect her interests. He borrowed against her home (now in his name) and vanished. Her children contacted me when the house was in foreclosure. There was not a good outcome to this case; the man had vanished and the loan was valid and the woman lost the home – she was not incompetent when she signed over her home, just over trusting. In another case I was more successful in recovering most of a man's life savings that he had transferred to his caregiver who promised lifetime care in exchange – there was no binding contract for such a lifetime of care and the man would probably have had to live another 150 years to get adequate value for the funds advanced. This man really did not know what he was signing and was clearly exploited by the caregiver. Fortunately, most of the money was returned.

The need for both protection and advocacy helped give rise to this new and special area of law for elders. In 1979, the American Bar Association established the Commission on the Legal Problems of the Elderly (now renamed the ABA Commission on Law and Aging). This multidisciplinary appointed commission, along with its professional staff, gathered together geriatricians, academics, advocates, members of the National Senior Citizens Law Center (a legal services program for the indigent), lawyers, judges, social work professionals, physicians, sociologists, political scientists and others to collaborate in identifying, defining and finding ways to address the special legal issues of aging. This, of course, has been most timely in anticipation of the largest cohort in world history, the Baby Boomers, approaching retirement age (the first Boomer applied for Social Security in January of 2008 at age 62). The Commission and its staff advocated for better options for surrogate decisionmaking (powers of attorney), advance directives (living wills) and alternative dispute resolution to avoid the courts. The Commission arranges studies of aging issues and has given many grants to assist incipient and promising programs throughout the US. In addition, it worked with AARP to develop the Legal Counsel for the Elderly. I served as a commissioner for several years in the 1980s.

In part the stimulus for all of these actions was the Older Americans' Act of 1965 which created the US Administration on Aging originally to address issues of nutrition, socialization and housing of older persons. This act was broadened over the years and ultimately included legal services as one of the priority services, provided generally through legal aid under the Legal Services Corporation. The Older Americans' Act, part of the Great Society programs of the 1960s, greatly raised the consciousness of the nation to the issues facing an aging population of increasingly isolated, underserved and poveritized seniors.

Until the mid-1980s, however, there was no private component of the legal services offerings to elders. In 1984, the ABA Commission on the Legal Problems of the Elderly sponsored a panel presentation at the ABA annual meeting in Chicago entitled "Doing Well by Doing Good." At this program, five lawyers from various parts of the United States who had been identified as providing legal services to primarily older clients and their families presented information about their practices, their clients and the types of work the lawyers did. The panelists were relatively stunned to learn that they were each doing similar work, each inventing a new area of practice in parallel. We met after the panel discussion and shared stories that were so similar that we were shocked. There was quite limited literature about law and aging and there was no organization to promote and develop the area of law. We made a commitment to try to locate other lawyers in the US doing similar work.

Approximately 70 such lawyers in private practice were located over the next year and a meeting was set for San Francisco, again underwritten by the ABA Commission. Of the 70 lawyers invited, more than 30 attended, funding their own expenses to attend. At this enthusiastic gathering, a decision was made to attempt to form an association of lawyers for the aging and the name National Academy of Elder Law Attorneys (NAELA) was decided upon. This was after the names National Association of Aging Lawyers and National Senior Lawyers were rejected. One attendee volunteered to incorporate the Academy as a non-profit corporation while others agreed to additional administrative tasks to create the organization. It was agreed that membership would be limited to those attending the San Francisco meeting while details were worked out.

By the efforts of these founders, NAELA was able to sponsor a national conference on elder law in Tucson, Arizona with attendance of more than 130 lawyers from around the country. This program served as an exchange of ideas, probably for the first time, concerning the legal issues of aging as related to the private practice of law (while many public interest lawyers were also present and addressed their analogous issues). Nearly simultaneously with the Tucson conference, the New York Times published a substantial article on this emerging area of law; the article listed my address as the interim office of the Academy - within a month, more than 15,000 inquiries were received from persons seeking local referrals to elder law attorneys. Many volunteers (mostly seniors) assisted in replying to this evidence of the need we were about to address and every letter received a response giving names of known practitioners in their areas.

Over the years, the Academy has grown to a professionally-managed organization that publishes a monthly newsmagazine, publishes a quarterly academic journal and organizes and presents several national educational conferences, symposia and institutes each year. There is a website: www.naela.org with substantial information about the practice. NAELA has professional advocates in Washington to assist in monitoring and advocating issues of its clients that come before Congress, service agencies and regulatory bodies. There are more than 5,000 NAELA members and there are more than 100 nationally certified Elder Law Specialist (Certified Elder Law Attorneys - CELA's), certified by the National Elder Law Foundation (created by NAELA) and approved by the American Bar Association as the certifying entity. See www.nelf.org for details on certified specialists.

The Academy has approved many state Chapters to consider and act upon issues on a more local level and has provided many other opportunities for the advancement of the legal issues and provision of services to the aging. All of the substantive issues addressed in the following chapters: Surrogate decisionmaking; Advance Planning; Powers of Attorney; Long term care and its financing; Public benefits; Private and public pensions; Right to treatment and right to refuse care; Protection of autonomy of seniors; Age discrimination; Housing issues and support services; Reverse mortgages; And the myriad other issues of law that confront the aging in special ways are issues that NAELA addresses in its publications, considers in its programs and tracks in the law-making bodies of the states and in Congress.

Finally, a key consideration of Elder Law Attorneys has been the issue of making their services accessible to older persons with courtesy and respect. Without ever assuming a client may be disabled, the Elder Law Attorney acknowledges that there may be some clients with special needs. Meeting these special needs with larger-print documents, available outreach and home visits, appropriate lighting and furniture, extraneous noise control and hearing assist devices are cornerstones of being able to serve our clients. Many elder law practices, like ours, have now become extended or holistic practices, providing legal services, care management services, financial management services and other assistance within their practices. Our clients have expressed great relief at finding all they need within one firm.

The issues of aging that we face as lawyers are, indeed, new ones as more people live longer and in new circumstances. The elder law attorney is in a unique position to help meet these needs. We are serving elders also when we have younger clients! I have a client in a distant state who contacted me concerning her mother who lives in Arizona. She was concerned when she had spoken to her mother, Sarah, recently that her mother seemed to be having memory issues. The daughter had concerns that her mother's housekeeper had been asking her mother for loans. The daughter had contacted a lawyer in her home state who had referred her to us. A care manager from my office discussed her mother's situation with her and then made a home visit to meet with Sarah and assess her needs. The care manager was able to determine that Sarah was able to meet her own daily needs and perform all of the basic activities of daily living including eating, dressing, getting into or out of a bed or chair, taking a bath or shower, and using the toilet and that she prepared her meals, did her own shopping, drove and used a telephone. What she also discovered was that Sarah, now 78, was not able to handle her finances. Sarah was widowed 3 years earlier and her late husband had handled all of the family's assets, made all the investments and paid all of the bills. The care manager explained that we had been asked to visit by the daughter, who was our client, and Sarah was quite forthcoming and welcoming of the idea of help with handling of her income, bills and investments. She acknowledged that she had been asked for loans by the housekeeper and "had made a few," and did not know the amounts. The loans were in cash from the cash machine and, no, she did not have any notes. We knew who our client was, i.e., the daughter, and advised Sarah of a possible conflict but Sarah was comfortable with our working with her. The care manager, well-versed in the options available, suggested that Sarah consider giving her daughter (or someone else) a financial power of attorney or that she consider setting up a trust to handle her finances. Sarah had not done any estate planning since well before her husband's death and had no powers of attorney in place. When Sarah met with one of our firm's lawyers, she could not adequately express her relief at finally turning over the management of her finances to her daughter; the daughter retained our firm to assist with the day to day bill-paying and investment decisions. The housekeeper was replaced and investigation of records showed that the loans had been minimal. Our elder law practice reaches out expressly to younger persons such as the daughter who have concerns about aging relatives, friends or neighbors.

Where does the profession go from here? There are many important areas that most elder law attorneys are not addressing in their practices. In the United States, much of the focus of elder law attorneys has been on issues of protecting the assets of persons who need long term care and assuring that clients qualify for Medicaid, the public assistance program that is intended for the poor. Interestingly, when I began practicing elder law (before it had that name) in 1975, Arizona did not have the U.S. Medicaid program (it joined the system in 1989). My practice did not include planning for long-term care expenses (other than divorce to divide assets, protecting about half for the spouse staying at home). My clients focused more on other issues such as protection of their autonomy, planning for possible disability, establishing or defending guardianships, estate planning for spouses or children with special needs, or addressing issues of abuse, neglect or exploitation. I was active in advocacy issues for older persons including universal appointment of counsel in guardianship proceedings, ending discrimination based on age and advocating greater accountability for persons handling money for older persons.

I was surprised then and continue to be surprised today at the considerable focus on asset protection – qualifying for Medicaid for nursing home expenses. This focus extends to individual practices, writing in elder law publications and a disproportionate number of programs on this topic at elder law continuing education programs. There is nothing wrong with financial planning for long-term care expenses, of course. Clients definitely need professional help with understanding the complex legal planning options open to them for getting help with the enormous expenses of such care as home care or skilled nursing home care. The United States is unusual in the developed world in requiring its citizens to reach poverty level before any help is available for such expenses. The elder law attorney who provides such counseling and assistance is providing a needed, ethical and valuable service. My firm does a significant amount of such counseling, helping clients through the morass of laws and regulations that provide for protection of resources. At the same time, NAELA continues to advocate for changes in the system that requires such planning, working toward universal health coverage which would include such care. Nonetheless, I think the future of the elder law practice requires lawyers to expand their skills to the other substantive issues of interest to older persons. This, I believe, is particularly important as the next generation of senior Americans, The Boomers, reaches age 60 and beyond – the first Boomers are, as of January, 2008, now applying for Social Security Retirement benefits at age 62. This is a different generation from those who lived through the Depression of the 1930s, a different generation from those lived and fought in World War II and a different generation from the cohort that came of age in the 1950s. The Boomers are the generation that matured in the turbulent '60s and '70s and share very different perspectives on life, retirement, saving and the entitlement to governmental programs. This is likely to be a much more demanding, much more litigious and much more activist group of older persons than their predecessors. At the same time, this generation is no longer just the "Sandwich Generation" caring for themselves and their children as well as their parents – they are becoming the "Club Sandwich Generation," caring also for their own grandchildren in many instances. This is the next great wave of our clients. Other countries have other generational/cohort issues and these need to be identified and addressed as the practice of law and aging matures in those different societies.

I believe that there will be far greater need for elder law attorneys to serve this generation and to be their advisors and advocates in many more areas than most elder law attorneys are now practicing. The future of elder law should include much more emphasis on issues of age discrimination in employment, in credit and in housing. Future elder law attorneys are also more frequently going to be looked to for help with managing finances and personal care, providing a more complete range of services for clients who have neither family nor friends to help with their daily needs. Recovery of assets for those who have been exploited will become more of our practice. Finally, the elder law attorneys should become more familiar with the laws concerning the abuse and neglect of elders in their own homes or in institutional settings – this becomes more important as greater numbers of elders are cared for by strangers. We need to be prepared to write and review care contracts, examine and recommend changes on institutional care agreements and be creative in developing options in a time when there will be more persons in need of care than there will be available caregivers.

At some point, it is likely or at least hoped that Medicare, the United States' national health insurance program for persons over 65 and certain disabled persons, will be expanded to include long-term care among its benefits. Should this happen, many of the elder law practitioners will see their practices rapidly dwindle. Yet, there is so much more to elder law that expansion of the training of elder law attorneys in these additional areas should see practices growing at the same pace as the growth of the population of older persons. The goal, I hope, will be to have the elder law attorneys provide comprehensive services in all of the areas of elder law, advocacy through national organizations of legal issues of aging and greater service to their clients.

I was in private practice for less than six months when I realized that providing the services my clients needed was going to require 24 hour/7 day availability if I was going to be able to actually agree to serve as a personal guardian or as agent under a power of attorney. While there are not many emergencies, there are some and I needed to be able to respond. I am not willing to be on call at all hours and, recognizing that a lot of what I was providing included social services, I decided to hire my first care manager. This began on a part-time basis but quickly became full time and ultimately led to needing more than one care manager on my staff. As I began handling more financial matters for clients under conservatorships, powers of attorney or as trustee, it became clear that someone with financial management skills was needed and I hired my first bookkeeper which position evolved into an accounting professional and then into a financial manager and then a full financial management department. Today, the firm I started in 1984 has six attorneys, two care managers, four financial management specialists and a large support staff. All of this evolved as the needs of the clients presented themselves. I think that this flexibility to identify and respond to the specific and changing legal and personal needs of older clients is going to be the key to the future of a successful elder law practice.

Chapter 2
Later Life Legal Planning

L.A. Frolik

2.1 Introduction

In response to client needs, elder law has expanded and is gradually redefining itself into "later life planning." While some still identify elder law with helping clients pay for long-term care, specifically in the United States by qualifying for Medicaid, the reality is that the practice of elder law is a rich mosaic of legal planning that is continually evolving to better meet clients' legal, financial and social needs and concerns.

While ten years ago, Medicaid planning to pay for long-term care was the focus of the practice of many, if not most, American elder law attorneys, that is no longer the case. Certainly Medicaid planning remains a core element in any elder law practice, but it is only one of many aspects of the practice that includes guardianship and mental capacity issues, long-term care planning, basic estate planning, drafting trusts, advising trustees, acting as trustees, creating special needs trusts, and advising clients as to their rights vis-a-vis assisted living facilities, nursing homes, and continuing care retirement communities. Elder law is even expanding into financial planning and "life care" planning that assist the client to address their financial and care needs of what for many is likely to be a very long life.

Though elder law attorneys have clients of all ages, because of the vicissitudes of aging, the very old have a particular need for legal assistance. The reality is that growing old, particularly growing very old, presents a host of legal problems. Consequently, for elder law attorneys, the very old, those age 80 or older, are becoming the foci of their practice.

2.2 Physical and Mental Decline and the Need for Legal Assistance

While the rate of aging varies, by age 80 most individuals have declined physically (Williams 1995, p. 3–41). They may have lost strength and flexibility, and so become referred to as frail. Other common, though not universal occurrences, include loss of hearing, loss of vision, often due to macular degeneration, and decline

I. Doron (ed.) *Theories on Law and Ageing: The Jurisprudence of Elder Law*,

in short-term memory. None of these makes the individual incapacitated, but alone or in combination they often result in older persons needing help with their personal or financial affairs.

While the law does not consider physical decline as reason for intervention or as something that precipitates the loss of rights, in reality significant physical decline makes an individual more dependent upon others. As individuals age, they lose physical vigor with the result that often they gradually (or suddenly as in the case of a stroke) can no longer handle their affairs alone and must rely on the assistance of others. Even the most commonplace of disabilities can render an individual incapable of handling some of their affairs. For example, the onset of macular degeneration can mean the loss of the ability to read, not just a newspaper, but more importantly the monthly bank statement or brokerage firm account. If the older person cannot read those statements, much less investment advice newsletters, the investor is no longer capable of actively managing his or her accounts. And the bills must be paid even if the older person's eyesight does not permit reading the mail. The answer, creating a durable power of attorney for property management is one solution, but just what authority the power should grant to the agent, who should be the agent, who should monitor the acts of the agent and whether the appointment of the agent will prevent the later appointment of a guardian are questions best answered by a knowledgeable attorney.

Beyond property management, a client's physical decline should also trigger a reexamination of the client's housing. As individuals reach their 80s, the effects of aging can begin to play a role in their choice of housing. The decline of physical strength and vigor is accompanied by a loss of muscle and joints becoming stiff and painful so that walking becomes more difficult and stairs can become an imposing obstacle. Because bones are weaker and more prone to breaking, slipping on a throw rug or on an icy front stoop creates a significant danger. Whereas formerly maintaining the house and lawn was merely an annoying chore, it now becomes a burden that saps the energy of the older person. Heavy cleaning and household maintenance become almost impossible with the result that the older house that an aging couple has lived in for many years may no longer be appropriate and may even create a risk of injury to them. Of course, the degree that physical decline affects an older person varies greatly from one individual to the next, but for many, the decline in physical capacity means that their current housing situation is no longer appropriate.

Loss of vision presents particular housing challenges for the elderly. With reduced vision, living alone or in a large house is not only difficult but unsafe. For example, hip fractures and other injuries resulting from falls often occur because impaired vision leads to tripping or missing a step. Almost all older individuals suffer some vision loss, including loss of ability to read fine print, sensitivity to glare, a decline of peripheral vision, and difficulty in adjusting from the light to the dark. In addition, many older persons experience eye diseases, for example, it is estimated that over 40% of persons age 75 or older have cataracts. For many, driving is no longer possible and even using simple tools, such as a screwdriver, or reading the instructions for a home appliance become almost impossible tasks, with the result that basic house maintenance – even replacing a light bulb – becomes increasingly difficult.

Hearing loss is also common among older persons. Beginning around age 50, most individuals experience a gradual loss of perception of the higher and lower frequencies, and one-third of adults between 65 and 74 and one-half of those between 74 and 79 experience presbycusis, a gradual deterioration of the inner ear that results in a permanent hearing loss. Although the condition can be somewhat corrected with hearing aids, diminished hearing makes individuals less sensitive to noise that could alert them to dangers or problems in the household. Just one more reason why living alone in an isolated or unsecured house may not be sensible for a very old individual.

Typically, as people age they also suffer a loss of short-term memory that is not associated with the onset of dementia. Rather a decline in short-term memory is a result of the natural aging of the brain and does not indicate a loss of mental cognition or capacity. The individual can reason as well as ever but can find it difficult to remember names, numbers, or specific information. As short-term memory fades, maintaining a house becomes more difficult. For example, trying to remember and compare the different estimates to paint a house can make it difficult to choose the best option. Remembering how to use a new household gadget, such as an automatic bread maker, can be frustrating. Of course, instructions can be written down and lists created of tasks to perform, but instructions can be confusing and lists lost.

While using notes and lists may solve common household puzzles, they are not as useful for more complex tasks such as managing finances where the ability to remember numbers, proportions, advice and warnings is so important. When memory fails, and if poor vision makes reading difficult, the natural response is to rely on someone else to advise, or even decide, as to how savings should be invested. Unfortunately turning to another for assistance can lead to bad advice, poor execution or even worse, exploitation and abuse.

The combination of reduced poor memory, physical vigor, declining vision, and hearing loss can make living alone or maintaining a house very difficult. Tasks that used to be simple, such as raking leaves, changing a fuse, or cleaning out gutters, can become almost insurmountable. Negotiating with individuals to come to the home and perform necessary maintenance or repairs can be a source of exploitation and even danger. Inappropriate housing is just one part of the mosaic of problems faced by the elderly that need to be resolved by a holistic approach that integrates solutions to social, personal and legal problems. An elder law attorney is uniquely situated to provide information and advice tailored to the client's needs. For example, any proposed change in housing requires an appreciation of the potential future needs of the clients so that today's "solution" does not become tomorrow's "problem."

2.3 Chronic Conditions and Housing Choices

Beyond the normal physical and mental declines of aging are the conditions and diseases that erode autonomy and lead to increasing dependence and vulnerability. Severe arthritis, pulmonary disease, strokes, and the most prevalent condition for the very old, dementia, all create the need for daily help (see generally Section 3 "Medical Conditions," in Beers 2004 [hereinafter Merck Manual]). The progressive nature of many chronic conditions means that over time the individual will lose physical and mental capacity and require ever increasing assistance.

It is at this point that many elderly individuals seek out an elder law attorney. What can a knowledgeable attorney do for such clients? Actually quite a lot. First is the matter of housing. Older individuals need to reside in appropriate housing, that is, housing that is safe, affordable and that presents maintenance and repair tasks that are proportionate to the ability of the older person's ability to deal with, either personally or by hire. Additionally the housing should be near enough to the services that the individual requires so that the services can be obtained at a reasonable cost and in a timely manner.

The client should be helped to triangulate his or her housing location, price and accessibility to services to determine whether the housing is appropriate now and in the future. Often the answer will be that the housing is either not currently meeting the needs of the client or that it will soon not be appropriate in light of the client's deteriorating condition. The solution is either in-home care or relocation. If the answer is in-home care, the attorney can explain the advantages, costs and potential disadvantages such as the problems caused by a careworker not showing up or the risk of theft from having a stranger in home. Relocation offers a variety of choices including an apartment or condominium located near needed services such as a hospital or physician; an age-restricted facility built to meet the needs of a physically impaired resident or that has therapeutic exercise facilities such as a swimming pool; supportive housing that provides housekeeping and meals; assisted living that provides daily personal care in a secure environment; assisted living that specializes in care of residents with dementia; a continuing care retirement community that will accommodate the increasing need for personal care; or a nursing home if the client's medical care needs require that level of care.

Spouses are the most common source of care for older persons with wives caring for husbands being the most frequent arrangement (Smith 2004, p. 351, 361). Because husbands are often older than their wives, have shorter life expectancies, and often are less skilled at or comfortable with being a caregiver, older women who need assistance either must find institutional or paid care or look to their families, usually adult children, for help (Lee et al. 1993, p. 9). Many adult children do provide the increasing care needed by an elderly parent, but while their intentions are good, they often realize neither the extent of the obligation that they are undertaking nor the accompanying legal issues that such an endeavor raises. What begins as an informal, ad hoc arrangement too often grows into a time consuming, stressful, demanding obligation.

When caregiving turns from running an occasional errand to providing regular personal care, the children would be wise to create a caregiving agreement that details the obligations of the children, who will provide the care and what compensation will be paid. Even if family members are willing to donate their labor, someone will have to pay for supplies and some care will likely be provided by specialists, such as visiting nurses. The need for a carefully drawn up contract to protect the rights of the older person is apparent to any attorney. Unfortunately, too often families think that because they are families, no legal advice is needed. They go ahead and agree that the house of the older person should "go to" the granddaughter who agrees to move in with her demented grandmother, not considering what should happen if the grandmother dies unexpectedly or if she should have to move into a nursing home in a few weeks because the dementia progresses rapidly. If the older person has the necessary capacity to change the deed, the family will often encourage him or her to put the house in joint name with the caregiver relative, who later decides the job is too difficult and moves to another state. Or the caregiver falls in love and has her boyfriend move into the house of the older person. The possible scenarios that could lead to bad outcomes are endless, which is the reason for legal advice and a contract.

If the decision is to hire professional assistance, the attorney will advise the family to use an agency rather than directly hiring a caregiver. Most elder law attorneys can direct families to professionals, not just caregiver agencies. Providing guidance and direction to a client or family in a time of stress and uncertainty is sometimes the most valuable "work" that the attorney can perform. By the time clients or families come to an elder law attorney, it is often too late to plan ahead, rather the need is for immediate action to meet a crisis. A good elder law attorney will have a list of potential housing options and service providers to assist the client or family in finding an appropriate solution.

A decline in mental capacity not only suggests the need to reexamine the client's housing, it also means that the client may now need or may soon need a substitute decision maker. The traditional answer to the loss of mental capacity is guardianship whereby someone petitions a court to find the older person incompetent and in need of a guardian. If the older person is found to be legally incompetent, the court may appoint a guardian who will act as a substitute decision maker and handle the older person's financial affairs and possibly also make decisions concerning the individual's personal affairs. Though filing for a formal guardianship is always a possibility, it is expensive, open to public scrutiny and subject to what can be unwanted judicial oversight and intrusion (for a discussion of guardianship, see Frolik 1981, p. 599). Consequently, most elder law attorneys believe it should be avoided if possible. Fortunately, with foresight and planning, guardianship can be avoided by the use of durable powers of attorney, trusts, and joint ownership arrangements.

2.4 Property Management Options

Even if the older person has lost some capacity, he or she may still have the mental capability to engage in planning for property management. Of course, if the older person has a spouse, the problem is greatly simplified. Typically most, if not all, of the assets are in joint name and so the well spouse can manage the couple's financial affairs. If the assets are not jointly titled, this might be the time to make that occur. If there is no spouse, one option is to jointly title the assets of the older person with an adult child that will permit the child to pay bills and manage investments. Care must be taken, however, to not inadvertently create joint ownership with rights of survivorship, which would make take the property out of the estate of the older person and make it not subject to the provisions of the older person's will. Jointly held bank accounts can be opened as "convenience" accounts so that the joint owner, the child, does not become the owner of the bank account at the death of the older person. A bank may even permit the child to sign checks on an account without becoming a joint owner.

Because of the post-death ownership problems caused by joint accounts with nonspouses, however, most elder law attorneys discourage their use. Instead they advise the client to sign a durable power of attorney that names an agent to manage the financial affairs of the older person. The older person can sign a durable power of attorney so long as he or she has mental capacity to sign a contract, meaning that the individual can reasonably understand the nature and effect of the act (Frolik and Radford 2006, p. 303, 313). This relatively low level of capacity enables even somewhat demented elderly to sign a durable power of attorney, and certainly elder law attorneys, recognizing the value of doing so, aggressively act to have clients sign such powers even when the client has diminished capacity. Of course, a durable power of attorney must be drafted to comply the legal requirements of the jurisdiction in which it is executed, but standard practice is to have the document witnessed and notarized. Typically more than one copy is signed because the agent acting under the power may have to present an original copy to a third party such as a bank or a brokerage house, who may even insist that they retain an original copy of the power.

The authority granted an agent under a durable power of attorney can be tailored to the needs of the person signing the power, who is known as the principal. Most very old clients are best served by giving the agent extensive powers, usually as broad as legally possible, in order that the agent has the flexibility to respond to whatever may arise. To ensure that the agent acts responsibly, some powers require the agent to account or report to a third party as to what the agent has done. For example, if one child, Marie, who lives near the older person is named as agent, she will have to make monthly reports to the other child, Gwen, who lives far away. This way Gwen will understand what Marie is doing and, if she has any concerns, she can raise them with Marie in a timely fashion. Many powers limit the amount of gifts to prevent the agent from dissipating the estate of the principal. Some powers permit gifts, but only up to a limited amount or only with the approval of another party. Many elder law attorneys, while recognizing the potential danger of permitting gifts by an agent, nevertheless believe that the durable power of attorney should permit them, albeit with some level of protection, to ensure that the agent will have the authority to engage in transfers as part of Medicaid or estate planning.

While the creation of a durable power of attorney for an older client is almost standard practice, it is often not sufficient. The potential problem of managing the finances of a mentally impaired older client is a major concern for elder law attorneys. As life expectancy lengthens and older clients own more assets, the need for more sophisticated ways of providing assistance for asset management and protection dramatically increases. After the use of a durable power of attorney, the next most popular solution is a revocable living trust in which the clients place their assets for protection in the event that they should lose metal capacity. The clients create a revocable living (*inter vivos*) trust and name themselves as trustees. (A couple often creates a single joint trust and names themselves as co-trustees.) They typically name their spouse, or is there none, the children, as the successor trustee or trustees in the event that they lose mental capacity or choose to resign if they believe that they are no longer effective at managing their financial affairs. At their death, the trust assets will be distributed as directed in the trust, which will provide for distributions that are consistent with the estate plan of the person who created the trust.

Many attorneys contend that older clients of means should routinely execute both a durable power of attorney and a revocable living trust. While a durable power of attorney is very useful for routine activities such as paying bills, a trust is preferable for asset preservation because of the extensive statutory and case law that governs the law of trusts and the obligations of trustees, as opposed to the less well developed legal obligations of an agent acting under a durable power of attorney. Third parties such as banks are also more willing to deal with a trustee than an agent. If the bank has a copy of the trust and a letter of resignation by the settlor of the trust as trustee, the bank should have no problem accepting the authority of the successor trustee.

If the older person has both a durable power of attorney and a living trust, the agent acting under the durable power of attorney will handle routine financial matters such as paying the bills, depositing checks, paying workmen and handling the checking account. The trustees of the living trust will handle the investment decisions and make any distributions or gifts beyond modest holiday gifts. The income of the older person would be paid into the trust with the trustees providing money as needed to the agent acting under the durable power of attorney.

2.5 Later Life Health Care Decision Making

The very old often face critical health care decisions including end of life decisions. The doctrine of informed consent requires that a patient direct his or her medical care and thus preserve personal autonomy. To give informed consent, the patient must understand his or her medical condition, the nature of the proposed treatment or procedure, the risks and benefits of the proposed course of action, and the alternative treatment choices. Clearly, informed consent demands a fairly competent patient who can understand and appreciate the information presented and the choices that must be made. Not all older individuals have that degree of capacity. Consequently, a significant aspect of elder law is to assist older clients to create documents that provide an alternative form of health care decision making in the event that the client is unable to give informed consent to health care.

The client can either sign a living will or appoint a substitute health care decision maker. A living will is a document that attempts to govern the end of life health care of a mentally incapacitated patient. Every state has a statute that grants individuals the power to sign a living will and imposes a duty on those treating the individual to respect and follow the directions in the living will. Though very popular a few years ago, and still in general use, most elder law attorneys are not enamored with a living will and prefer that the client appoint a surrogate health care decision maker. Living wills are considered to be too inflexible and not finely tuned enough because no one can anticipate what medical decisions need to be made near the end of life (Perkins 2007, p. 5). While a living will can indicate that the individual prefers death to extraordinary medical care, that degree of generality may not be much help when it comes to making specific treatment decisions. Instead, elder law attorneys now prefer to have the individual sign a durable power of attorney for health care, or as it also known, the appointment of a surrogate health care surrogate decision maker, to whom the treating physician can turn to discuss treatment options if the patient has lost the capacity to participate. The agent or surrogate will be empowered to act on behalf of the patient and have authority to act, just as if he or she were the patient, including the right to terminate life sustaining treatment.

Of course, merely appointing a surrogate decision maker is not the end of the matter; the older person must talk to the surrogate about the kind of treatment decisions to make. Ideally the surrogate would understand the values and hopes of the principal and translate that knowledge into decisions that respect and promote the views of the principal. Elder law attorneys realize that the topic of how to be treated at the end of life is something that many would prefer to ignore or put off to a later day. Getting clients to appoint a surrogate and to discuss thoughtfully end of life care with that surrogate is an important aspect of counseling older clients. As the client ages, the document should be revisited, as experience teaches that the health care decisions that make sense for a 70-year-old may not be appropriate for a 90-year-old. The surrogate should be counseled to be aware of the possibility that the older person could someday suffer from depression, dementia or delirium with the result that the older person might resist medical care but appear to have capacity. In such a situation the agent must be prepared to intercede and insist that the older person be aggressively treated for the condition that is interfering with his or her ability to make reasonable decisions about health care.

While planning for a surrogate to make acute health care decisions is important, planning for chronic health care is also a compelling need. Chronic care needs arise from both physical, such as strokes, and mental, such as dementia, causes. Planning for chronic care means thinking about how to obtain necessary care and how to pay for it. As discussed, an individual's choice of housing is an important component of chronic care. For the vast majority of the elderly, the question is whether adequate care can be provided at home or whether they will have to move into an assisted living facility or nursing home. Although having care provided at home is the overwhelming preference of most elderly, too often it is not possible. While a spouse can often provide care for a time, if the chronic condition continues to progress and the patient's needs increase, professional assistance may be needed, and home care may no longer be affordable or in some cases even feasible.

2.6 Planning for Chronic Care Needs

Because long-term care planning is a critical element of later life, elder law attorneys often serve as advisors to spouses and families as to the practical choices open to them when a spouse or parent has chronic care needs. Any elder law attorney can tell of numerous phone calls from adult children who relate how Mom or Dad, "Just can't live alone any more." The family or spouse needs a realistic discussion of how to provide for the needs of the older person—realistic both in terms of quality of care and of cost. For a family, caring for someone with a progressive chronic condition may be a unique experience, but for the attorney, the story is a familiar one. The attorney can advise as to how to find and pay for home health care, respite care and, if necessary, institutional care.

For individuals with in-home care needs, a well spouse is sometimes capable of handling the situation, but often he or she cannot. If, for example, a husband, age 88, has suffered a stroke and requires assistance with dressing and bathing, his spouse, age 85, may not have the strength to act as his personal care assistant or she may become worn down or exposed to injury, such as falling, while assisting her husband. Couples need to realize that caregiving can be very difficult both emotionally and physically, and that a little help can go a long way to preserving the health of the well spouse. An elder law attorney can often give advice as to a way that permits the well spouse to hire help while avoiding feelings of guilt for needing assistance. The mere existence of a spouse, however, does not mean that he or she will be capable of doing much for the spouse who needs long-term care. Often the "well" spouse has needs of her own. For example, if she has macular degeneration and cannot drive, she will have difficulty in shopping and running errands, and her poor eyesight may make it very difficult to manage the couple's finances.

Individuals with chronic care needs must appreciate that their care needs are likely to progress along a continuum, beginning with rather modest assistance to ever increasing levels of care. They must also understand that their care can be divided roughly into two types: custodial and medical. Custodial care refers to non-medical, personal care of the individual such as bathing or dressing. Medical care refers to the provision of medical services such as pain management or physical therapy as well as treatment for the underlying condition.

Custodial care is needed by individuals who have lost the ability to independently perform day to day care needs, which gerontologists have classed as activities of daily living (ADLs) and instrumental activities of daily living (IADLs). ADLs are essential self-care activities and are usually listed as eating, dressing, bathing, transferring between a bed and a chair, and using the toilet (Frolik 2006, p. 65). The inability to perform one of these activities is called a "deficit." Individuals with deficits in ADLs cannot safely live alone because they need daily care and are typically candidates for residence in an assisted living facility where they will receive help as needed. For example, the facility will have special step-in bathtubs that may enable the resident to safely take a bath alone or the facility may provide an aide to assist the individual to take a bath and to get dressed.

IADLs are not as rigidly defined as ADLs and include common activities in and out of the home such as shopping for groceries, preparing meals, performing housework, taking drugs as instructed, going on errands, using a telephone, and managing basic finances such as paying the monthly utility bills (see Section 3 "Medical Conditions," in Merck Manual 2004, p. 22). Individuals with IADL deficits can often continue to live at home, be it a house, condominium or apartment, though they do need assistance, which can come in several forms. An elder law attorney can explain the options and help craft an affordable and safe solution. Groceries can be delivered, taxis or publicly sponsored transportation for the elderly can solve the problem of getting around town, a relative with a durable power of attorney can look over the mail and pay bills, and a daily attendant can visit the older person for a couple of hours each day to assist with housework, prepare a meal and see that the older person takes his or her prescribed medication.

The ability to drive is critical for most Americans to be able to perform the IADLs as there is rarely adequate public transportation, and what they need is usually not within walking distance. Here again, the elder law attorney can raise the issue of whether it is realistic to expect a very old spouse to single handedly perform all the IADLs needs of the couple. Merely asking the well spouse to do that much driving, in light of her slower reflexes and poor eyesight, may put her at considerable risk. Because the burden and risk on the well spouse may not be appreciated by her or even by the adult children, it is part of the duty of the elderly law attorney to raise the issue and point out that the couple may need some assistance.

If no one is available to volunteer to help, the elder law attorney can assist in locating paid help such as a driver, a personal assistant or a daily homemaker assistant. A better solution in many instances is to hire a professional geriatric care manager to either perform the work herself or coordinate the hiring and supervision of others. By virtue of their training as a registered nurse, licensed social worker, or gerontologist, geriatric case managers understand the physical, mental and social needs of the elderly, can assess the client's capabilities and needs, judge what is a realistic expectation of help from a spouse, and coordinate volunteer efforts of family and friends. Usually the geriatric care manager is an independent proprietor but a few elder law offices employ geriatric care managers as staffers and some hire the geriatric care manager as an independent contractor on a client by client basis.

The geriatric care manager knows what resources are available to meet the other needs of the older person and will create a plan for the provision of required services either by the geriatric care manager or by others at a lower cost. For example, the geriatric care manager might only monitor others to perform specific tasks, such as a part-time housekeeper who will do the grocery shopping and prepare healthy meals for the older person. If the older person has no spouse and no available child, the geriatric care manager could drive the individual to doctor's appointments and discuss the care plan with the doctor to assure that it is followed and that all drugs are taken as directed.

Assuming that a geriatric care manager is hired, the role of the elder law attorney is to draft of review the contract between the geriatric care manager and the client (who could be the individual or a family member) and ensure the contract properly deals with the variety of contingencies that can occur. For example, what happens if the geriatric care manager becomes ill and unable to perform the promised services? What happens if the client loses confidence in the geriatric social worker? A good contract will be adaptable to whatever should occur. The attorney will also review the insurance carried by the geriatric care manager to see that it will provide adequate financial protection in case the geriatric care manager or someone employed by the care manager negligently injures the older person or financially exploits her. The attorney will also advise the client to watch out for self-dealing by the geriatric social worker, hiring of help based on nepotism or kickbacks rather than merit, and fraudulent billing. The client must trust the geriatric social worker, otherwise why hire her, but that trust should be tempered with the reality that it is those whom older persons trust who can most easily abuse and exploit them.

In lieu of hiring a geriatric social worker, some single older clients will prefer to hire a family member, usually an unmarried adult child, to care for them. Often the younger caregiver moves in with the older person, and agrees to provide daily custodial care in exchange for a weekly wage, a lump-sum payment or is paid by being given title to the house. All aspects of such an arrangement demand the need for advice from an attorney. While the client may believe that because a child is providing the care, no formal contract is needed, experience with these arrangements has taught the elder law attorney otherwise. Absent a written contract, the parties may have quite different expectations as to the role of the child. The older person may expect the child to perform many more services and increase the amount of assistance as the needs of the older person become greater. In contrast the child may expect to help for only a few hours a week and never anticipate providing daily care of a very personal nature such as helping the older person dress or bathe.

A contract between the parties will make clear the amount and type of care expected to be performed by the child, and when and how often the care is to be provided. For example, if the child is supposed to perform as a driver for the older person, is that service a daily event, once a week, on demand by the older person or only when convenient for the child? The contract should provide vacation time for the child and provide that the older person will move into respite care for a specified period of time, which might mean moving in with another child or into an assisted living facility. The contract should detail what happens if the care needs of the older person become too demanding or too much of a medical nature for the child to undertake. If the parties cannot agree that the responsibility has become too great, the contract should provide for a third party to decide whether the child is relieved of his or her care obligation. Finally, the contract should address how, when and who decides if the older person should be institutionalized in an assisted living facility or a nursing home. Here too a third party may have to make the call because the caregiver may be a bit too eager to be relieved of his or her duties and the older person too determined to remain at home. Of course, the third party cannot force the older person to move, but the third party could decide that the contract is terminated because the needs of the older person can no longer be met by the caregiver.

Most importantly, the contract must detail the compensation to be paid to the child and what conditions can justify withholding it. A popular method is a fixed weekly wage based upon the estimated value of the services to be performed. An hourly wage is possible, but most older persons do not want to have to gauge whether the value of service is worth the hourly cost. In some instances a lump sum is paid to the caregiver to pay for all future care. What looks like a gamble as to how long the care will be necessary, is in fact a first step in creating eligibility for Medicaid to pay for long-term care, which will pay for the cost of a nursing home if the resident has spent down his or her assets. By prepaying the child, the older person will have reduced his or her assets to nearly zero and live on a pension and Social Security benefits. When the older person can no longer be cared for at home, he or she will move into a nursing home and apply for Medicaid. If the older person had given away assets to the child, Medicaid would have imposed a period of ineligibility. By purchasing care from the child, the older person can transfer assets to the child but not be denied Medicaid on account of having made gifts (Frolik and Brown 2007, pp. 14–45). Of course, the amount of the lump sum payment to the child must be reasonable in light of the probable care needs of the older person, but the care need not end when the old person enters the nursing home as the child can continue to provide modest amounts of supplemental care and supervision to the older person.

2.7 Long-Term Care Insurance

Paying for long-term care in an institution poses significant hurdles. With the annual cost of a nursing home in the United States ranging from $60,000 to $120,000 a year, many clients consider purchasing long-term care insurance. While a few elder law attorneys have become licensed to sell such insurance, most merely advise their clients about the nature of the product and explain its advantages and disadvantages.

Long-term care insurance, issued by private insurance companies, guarantees a fixed monthly cash payment in the event that the insured's health meets the criteria that triggers the payment of the policy benefits (Frolik and Brown 2007, pp. 15–33). Almost all long-term care insurance policies are "indemnity" policies that pay a fixed dollar amount each day that the insured qualifies for benefits. For an increase in the premium, most policies offer inflation protection by increasing the daily payment by a set percentage each year, because the benefits, which seem adequate when the policy was taken out, may prove woefully inadequate when finally paid ten or twenty years later.

Long-term care insurance policies usually pay benefits for skilled nursing home care, custodial care (personal care) such as is provided by an assisted living facility, and home care if the individual meets the criteria as stated in the policy. Most policies are sold to individuals between ages 50 and 84, although a few companies sell to those who are younger and a handful sell to those that are older. The annual premium naturally rises with the age at which the policy is purchased, though once purchased, the policy premium will not be raised unless the company raises premiums for all similar policies, which unfortunately because of inflation is very likely. Daily benefit amounts vary from policy to policy and the number of days of coverage can vary from one year to life. Part of the role of the elder law attorney is to help the client determine just what combination of benefits is needed. Some want very high daily benefits, others are more concerned that benefits can be paid indefinitely.

Typically the payment of benefits is triggered by:

- An inability to perform a specified number of "activities of daily living,"
- The necessity for supervision and care because of cognitive impairments, or
- The need for long-term care because of a medical necessity, such as a stroke.

A common feature of long-term care insurance policies is a deductible period (also called an elimination period) that requires the policyholder to pay for a specified period before benefits are paid. The longer the deductible or elimination period, the lower the premium. The elder law attorney can point out to the client that it is financially sensible for the insured to pay for the first weeks of care and so realize significant savings in premiums over the life of the policy. The purpose of insurance is to protect against unacceptable risks, and most clients will find that the risk of paying for at least six months of care, while not welcome, is acceptable. Some policies have an accumulation period to satisfy the elimination period. That is, they specify a period during which the total number of deductible days, even if they are not consecutive, can be counted towards meeting the deductible. Accumulation periods are typically three times as long as the elimination period. Some policies have only one (lifetime) elimination period; others restart the deductible period if the days spent in the nursing home are separated by a certain period, such as six months or one year.

Unfortunately many older persons do not think to purchase long-term care insurance until it is too late to do so. Policies are generally not sold to those age 80 or older, and the annual premium rises sharply past age 75. Even younger potential purchasers may have to accept a delay in the starting date of benefit payments because of a preexisting condition. While most policies do not absolutely bar payment for preexisting conditions, most will not pay for the care that arises from a preexisting condition until the individual has qualified for benefits for at least six months. The lesson is that the time to buy long-term care insurance is long before the need for it arises.

The elder law attorney must point out to the client that buying long-term care insurance can be long-term commitment. For example, a sixty-year-old who buys a policy may have to keep paying premiums for twenty or thirty years. Even then the insured may die and have never qualified for the payment of benefits. Because of the need to keep paying premiums for many years, the temptation to let the policy lapse is very great, particularly if the premiums rise and the insured is pressed to pay other expenses. Letting a policy lapse is so unadvisable that the attorney may advise adult children that they may wish to pay for their parents' long-term care insurance if the parents can no longer afford to do so. That payment should be seen as an investment to protect their potential inheritance from being sharply diminished by long-term care costs.

Before buying long-term care insurance, clients should carefully consider whether that is the wisest course of action. An alternative is to move into a continuing care retirement community (CCRC) that promises to provide appropriate care at all levels of need including nursing home care. The large down payment required to enter the home, which may not be refundable, represents prepayment of possible custodial and medical care expenses. The trade-off is the assurance of quality long-term care at the price of the monthly fee of the CCRC, which, of course, will rise over time. Still, the resident of a CCRC knows that appropriate care will be available whatever his or her future care needs may be. As long as the client can afford the relatively high monthly payment, a CCRC may be an attractive alternative to the purchase of long-term care insurance.

Many elder law attorneys counsel some of their clients to forego long-term care insurance and to expect to rely instead on Medicaid. Clients with modest savings and moderate income often conclude that the cost of the insurance is prohibitive in relation to the potential benefits. For example, if a single individual has $300,000 in savings, he or she can afford three or four years of long-term care at a cost of $70,000 to $100,000 per year. If the care needs extend longer than that, the individual will have exhausted his or her savings to be sure, but Medicaid will then pick up the cost of the care. Conversely, clients with a high net worth can self-insure their long-term care costs. For example, a couple with $3 million in savings plus annual Social Security income of $35,000 can absorb as much as $1 million in long-term care costs (about ten years worth) without unduly affecting the well spouse since the income produced by the remaining $2 million should be at least $80,000, which when added to the Social Security payments would produce over $100,000 a year income for the well spouse.

When considering whether to buy long-term care insurance, the client needs to describe the specific risk that is feared because insurance should only be purchased to protect against unacceptable losses. For example, a couple has $400,000 in savings. They buy long-term care insurance that covers both of them because they want to protect the standard of living of the well spouse should the other have to move into a nursing home. Another example, a widowed older woman with $500,000 in savings buys long-term care insurance to help insure that upon her death, her estate will be able to adequately fund a trust for the benefit of her develop mentally disabled adult son. Finally, a couple with $800,000 in savings buys long-term care insurance both to protect the financial well-being of the well spouse and to protect the value of their estate that they wish to leave for the benefit of their three grandchildren whose mother, their daughter, recently died of cancer at the age of 45.

2.8 Counseling a Client with Diminished Capacity

A very significant part of elder law is dementia planning, which is practically a sub-specialty in itself. Almost always begun after the older person has begun to show significant signs of the disease, it requires property management planning including the creation of durable powers of attorney and trusts, long-term care arrangements, and planning for surrogate health care decision making. The lawyer becomes the guide who helps the family arrange an affordable, safe living arrangement for the older person by assisting the family to select from the array of possible solutions including in-home assistance, a continuing care retirement community, assisted living or a nursing home and also helping the family to craft a method of paying for the needed long-term care. And of course, if a move to a nursing home is indicated, the elder law attorney can investigate the possibility of planning for Medicaid eligibility.

When dealing with older clients, the elder law attorney knows that certain techniques help make a meeting a success. They include:

- Keeping meetings short.
- Meeting at the home of a client who is excessively fatigued or confused by meeting at the attorney's office.
- Using a round table so that the planner can sit next to the client.
- Using large fonts on all printed material to be shown to the client.
- Making sure that there is no background noise.
- Removing unnecessary papers from the meeting table.
- Not sitting in front of a window, which makes it difficult for a client to read the expression on the planner's face.
- Not accepting phone calls during the meeting.
- Taking frequent breaks.

If the client has sufficient capacity, the attorney can help draft an estate plan. Later life wills are a common client need. While the attorney-client relationship may have arisen from other legal needs, often the client also decides to write a new will. While there is no exact information as to the average age of estate planning clients or the age at which individuals write wills, anecdotal evidence indicates that some very old clients, often surviving spouses, revisit wills made in their younger years. Unfortunately, too often it is the spur of declining mental capacity that moves the client to finally update his or her estate plan.

Before the estate plan can be crafted, however, the client and the attorney must work out a plan for paying for potential long-term care. The client must consider what an estate plan should look like in the event that the estate has been depleted by thousands of dollars spent on long-term care or the attorney should try to transfer assets during the life of the client in order to qualify the client for Medicaid payment of long-term care costs. While it is now very difficult to transfer assets and still qualify for Medicaid, there are options, including an irrevocable trust with all income paid to the settlor (the client). The transfer of assets into the trust does create a five year period of possible ineligibility for Medicaid, but careful drafting of the trust distribution provisions can minimize the problem.[1]

---

[1] Medicaid planning is complicated and should not be attempted without the advice of a qualified elder law attorney. See Frolik and Kaplan 2006, p. 129.

2.9 Late Life Estate Planning

Wills written by those past age 80 or 85 usually represent the last will that the individual will execute. Because very old clients typically do not expect to ever write another will, they take great pains to get this one right. The lawyer who has such a client is used to the special aspects that distinguish this exercise in estate planning from those of younger clients. At a minimum the lawyer feels the press of time. There is a need for timely resolution and execution of the plan because the death or mental incapacity of the client is a very real possibility. Yet a quick resolution may be hampered by a client whose hearing or eyesight makes it a challenge to communicate or whose diminished mental capacity makes it difficult to proceed with any haste. Other very old clients whose minds are intact nevertheless delay making decisions because they are uncertain of what is the right thing to do. The very fact that this may be their final estate plan may cause them to hesitate to act.

The common problem of diminished short-term memory is also a stumbling block because the client who has difficulty recalling information finds it hard to make essential decisions. For example, the proposed plan involves generation skipping, but the day after the meeting with the attorney, the client is very confused as to why that technique is being employed. Declines in hearing and vision also complicate estate planning because of the difficulty of communicating complex proposals to the client. Other possible physical limitations include a loss of energy and attention during a long meeting, different levels of mental capability between spouses, and a client's insistence on having a third party present at meetings or, even worse, a client who will not act until the plan is approved by a third party such as a child or a friend.

An older client may be often uncomfortable or confused by financial or estate planning. The client may be frail, easily tired, and unsure about the complex matters being discussed, such as avoidance of the federal estate tax. Careful counseling by the attorney can usually help the client feel comfortable about these matters. Some very old clients have views towards their property that can cause difficulties, such as being fixated on the distribution of the personally or heirlooms while ignoring the distribution of the intangible, but substantial, assets. A few older clients become agitated when faced with the limitations and failures of their heirs who themselves are not young. For example, an older client may be upset with the course of the lives of adult children or may not approve of the lifestyles of grandchildren. Very old clients typically have descendants whose marriages, divorces, remarriages, life partners and children in and out of marriage may confuse the client as to what is an appropriate manner distribution of the estate. Dealing with these emotional issues is not easy, but solutions must be found if the estate plan and other important documents, such as the appointment of a surrogate health care decision maker, are to be completed.

When drafting an estate plan for older clients, the attorney must pay close attention to whether the client has testamentary capacity. Fortunately, in the United

States, the law favors a finding of sufficient testamentary capacity. As described by a Colorado court:

> … a person has testamentary capacity when the person (1) understands the nature of the act, (2) knows the extent of his or her property, (3) understands the proposed testamentary disposition, and (4) knows the natural objects of his or her bounty, and (5) the will represents the person's wishes. An individual lacks testamentary capacity under the insane delusion test when he or she suffers from an insane delusion that materially affects the disposition of the will.[2]

Because testamentary capacity is so low in the United States, less than what is needed to engage in a valid contract, for example, usually the attorney will decide that the client has sufficient capacity to execute a valid will. Sometimes the attorney will rely on the doctrine of the "lucid interval," which refers to an interval of apparent mental clarity (or at least capacity) for an otherwise incapacitated individual. A will executed during a lucid interval is valid despite evidence even if the testator seemed otherwise incapacitated on the day that the will was signed. The doctrine often has application with a testator with dementia whose mental capacity varies from hour to hour. For example, many elderly suffer from "sun-downing." They have capacity in the morning, but gradually become more disoriented and confused as the day progresses. If the will is signed early in the day, the client may have testamentary capacity, though not have it later in the day as the level of dementia rises.

If the attorney is uncertain that the client has capacity or fears a potential challenge to the will, the attorney should arrange for an examination of the client by an appropriate professional. Some planners videotape the signing of the will to create a record of the client's mental capacity, but to do so acts as a red flag and raises the question as to why it was thought necessary. The better solution is to have the client's mental capacity evaluated by a qualified professional such as a geriatric psychiatrist.

Frequently very old clients insist that a third person – other than the spouse – be present when they meet with the attorney. Such a request raises concerns as to confidentiality, possible conflicts of interest and undue influence. Of course a client can waive confidentiality by signing a waiver, but the attorney should also meet alone with the client and inquire why the client wants the third party present and whether any subjects should be off-limits in front of that individual. During the private session with the client, the planner should discuss any possible conflicts of interest and probe to see if there are any grounds for suspecting undue influence.

Because it is very difficult to successfully challenge the validity of a will based on a claim that the testator lacked testamentary capacity, those who wish to overturn a will often claim undue influence (Frolik 1996, p. 841). The older the client, the more likely an allegation of undue influence, which invalidates a will or a bequest on the basis that the will or bequest reflects the wishes of the agent of the undue influence rather than those of the testator. Even though the testator had the necessary capacity to execute a will, it can be invalidated if the undue influence overrode the will of the testator.

---

[2] *In re* Estate of Romero, 126 P.3d 228 (Colo. App. 2005).

The elements of undue influence are:

- A confidential relationship existed between the testator and the influencer.
- The influencer used that relationship to secure a change in how the testator distributed his or her estate.
- The change in the estate plan was unconscionable or did not reflect the true desires of the testator.
- The testator was susceptible to being influenced (Frolik 1996, p. 850).

Confidential relationships include a planner-client, clergy-congregational member, doctor-patient, nurse–patient, parent–child, adult lovers, collateral relatives, siblings, housekeepers and even friendly neighbors. If the older the client was emotionally dependent upon another person, that dependency can be interpreted as susceptibility to undue influence (Frolik 1996; p. 841). All that is necessary is that there be a special trust and confidence between the testator and the alleged influencer. The distribution scheme is both the motivation and the grounds for a finding of undue influence (Frolik 1996, p. 841). For example, a will that favors a new "friend" and leaves nothing to the children is vulnerable to a finding of undue influence. The attorney also must be alert to a claim of undue influence if the older testator insists upon making substantial gifts to "helpful" neighbors or to a "caring" nurse.

The older or more frail the client, the easier it is to claim susceptibility to undue influence. To discourage such claims the attorney should have frank discussion with the client as to the reasons for the bequests, keep detailed notes as to the reasons for the plan of disposition and perhaps encourage some gifts to the "natural" heirs to discourage them from supporting a challenge to the will.

Attorneys must also be alert for signs of clinical depression, which is evidenced by depressed attitude, irritability, anxiety, lack of self-confidence, low self-esteem, poor concentration, poor memory, social withdrawal, hopelessness, and recurrent thoughts of death or suicide. If depression is suspected, the planner should urge the client to seek a professional evaluation because clinical depression can be alleviated and even cured.

2.10 Retirement Planning

Many elder law attorneys now consider retirement planning as part of the practice of later life legal assistance. It is characterized by financial planning and in the United States especially by planning with how to deal with pensions, §401(k) plans and Individual Retirement Accounts. Although most clients will have created a financial plan when they transitioned from work to retirement, many clients need to revisit those plans as they age. They need to consider the current wisdom of the their pattern of investments, consider whether they are spending too much or even too little, and revisit whether their financial plan is coordinated with their estate plan. Many clients prefer to consult financial planners for advice about these subjects, and many attorneys are reluctant to give financial advice. Some attorneys, however, have obtained the necessary licenses to engage in financial planning and some even sell products such as long-term care insurance. These attorneys believe that it only makes sense for them to expand their practices to meet the needs of their clients.

Other aspects of retirement planning include giving advice about long-term insurance, supplemental health care insurance, and housing and relocation choices. The attorney may also provide information about the tax consequences of the client's re-entry into the workforce as a part-time employee, consultant or entrepreneur.

Some elder law firms have become full service legal, financial and social service responders. They not only provide legal assistance, but employ or contract with other professionals such as accountants and geriatric care managers to provide the services required by the client. These firms have adopted a multi-disciplinary practice and provide a variety of services to meet the needs of their clients.

Some elder law firms are now entering into life care contracts with clients under which the firm agrees to provide lifetime provision of long-term care advice and supervision. The client pays a one time fee that is based upon the attorney's estimate of the value of the future services that the firm will have to provide the client. The law firm either employees or contracts with a nurse or geriatric care manager who helps evaluate the client's needs and if necessary provides services. For example, the attorney may help devise a retirement plan that includes provisions to pay for long-term care, which might include creating eligibility for Medicaid. The law firm will monitor the client's condition and see to it that the client is advised how to obtain the proper level of care in an appropriate environment. If the client enters a nursing home, the firm will monitor his or her condition and help assure that the client receives quality care.

As elder law expands its horizons to meet the needs of older clients, it becomes less strictly the practice of law and more the provision of later life assistance. As such, its future looks bright indeed.

References

Beers MH (ed.) (2004) The Merck manual of health and aging [cited Merck Manual]

Frolik LA (1981) Plenary guardianship: An analysis, a critique and a proposal for reform. Ariz L Rev 23:599

Frolik LA (2006) The law of later-life health care and decision making. p. 65

Frolik LA, Brown MC (2007), Advising the eldery or disabled client, 2nd edn. with supplements

Frolik LA, Radford MF (2006) "Sufficient" capacity: The contrasting capacity requirements for different documents. NAELA J 2:303

Frolik LA, Kaplan RL (2006) Elder law in a nutshell, 4th edn.

Frolik LA (1996) The biological roots of the undue influence doctrine: What's love got to do with it? U Pitt L Rev 57:841

Lee G et al (1993), Gender differences in parent care: Demographic factors and same-gender preferences. J Gerontol 48:9

Perkins HS (2007) Controlling death: The false promise of advance directives. Ann Internal Med 147:51
Smith PR (2004) Elder care, gender, and work: The work-family issue of the 21st century. Berkeley J. Emp. Lab. L.:25:35
Williams ME (1995) The American Geriatrics Society's complete guide to aging & health. Harmony Books, New York

Chapter 3
A Therapeutic Approach

M.B. Kapp

3.1 Introduction

Legislative and regulatory bodies around the globe have enacted over time an extensive *corpus* of law intended to improve the well-being of older persons and the quality of their lives. These statutes, regulations, and executive orders, as well as evolving common law doctrines enunciated by the courts, provide special protections or create distinct rights or privileges solely or at least partially on the basis of chronological age. Frequently, the older intended beneficiaries must become personally involved – voluntarily or involuntarily – with attorneys and the legal system to take advantage of the laws intended to serve their interests. Advocates for the older population (*see generally* Symposium 2004) generally make an assumption about the positive therapeutic efficacy for older persons of these laws and involvement in the legal system, but we seldom actually carefully examine the validity of that assumption after a law has been in place for a while or the elder's involvement in the legal process has taken place. Such careful examination – an assessment of the actual, rather than the theoretical, impact of the legal system on the older persons it is intended to serve, as well as their families and society as a whole – is the goal of the Therapeutic Jurisprudence (TJ) concept.

TJ suggests the law and legal practice have inevitable consequences for the well-being (psychological, physical, financial, and other) of involved people (see generally Wexler and Winick 1996 anthology of therapeutic jurisprudence essays). It is an interdisciplinary, interprofessional field of legal scholarship – a particular analytical lens – with a pragmatic, realist law reform agenda. TJ seeks to identify the therapeutic and antitherapeutic (and therefore counterproductive) effects of law and involvement in the legal process, and to shape the law and legal practice (Stolle et al. 2000) in ways that diminish the antitherapeutic consequences and maximize the therapeutic potential for actual, identifiable (as opposed to abstract) older individuals (see e.g., Wexler and Winick 1991).

Closely related to TJ is the concept of Preventative Law. "Preventative Law is a perspective on law practice that seeks to minimize and avoid legal disputes and to increase life opportunities through legal planning" (Stolle and Wexler 1997, pp. 25, 27). "It takes a proactive approach to a client's legal issues, i.e., the attorney and

I. Doron (ed.) *Theories on Law and Ageing: The Jurisprudence of Elder Law*,

client work together to determine the potential legal issues that clients may or will face and what can be done to plan for those issues in order to avoid undesired outcomes" (Faulkner 2005, p. 685, 687 (citing Stolle et al. 1997 p. 15,16)). Estate planning and health care planning (see Chap. 2) are two areas of elder law practice that could and should readily incorporate a combination TJ and Preventative Law approach to legal services. Both in the United States and elsewhere, there often is a significant gap between professional and public knowledge about preventive legal tools, on the one hand, and the actual use of those tools by older adults, on the other (Doron and Fal 2006 p. 4).

Also compatible with the TJ concept is the Jurisprudent Therapy Perspective.

> The perspective of jurisprudent therapy (JT) is an extension of therapeutic jurisprudence and serves as a means for assessing mental health science, practices, and roles with the goal of promoting principles of justice and human freedom while minimizing anti-jurisprudent outcomes for the client. Thus, JT provides an interdisciplinary approach aimed at developing and improving mental health science and forensic practice for individuals, as well as the general public (Marson et al. 2004, pp. 71, 72–73 (citing Drogin 2000 p. 489)).

Why should caregivers to, and advocates for, older persons who must function at ground level in the everyday world be concerned about a new analytical lens for evaluating specific laws that claim to benefit the elderly? There are, in fact, a number of good reasons for a TJ inquiry into the question of how well the positive therapeutic effects of specific laws and legal practices justify the faith underling our zeal to use governmental authority to regulate on behalf of older purported beneficiaries.

First, there is a recognition that performing oversight of, and enforcing compliance with, legal regulation and practice standards entails real financial and social costs, including costs to the intended beneficiaries in terms of access to, affordability of, and choice regarding needed services or resources. When we regulate something, we ordinarily get less of it and pay more for it. An honest, objective regulatory cost/benefits comparison should be a central part of the TJ analysis for any current or proposed legal rule or practice.

Second, public policy taking the form of directive, command-and-control regulation ("Thou shalts" and "Thou shalt nots") must be investigated, in terms of efficiency and effectiveness, in comparison to alternative methods of achieving the same laudable quality assurance and elder rights goals. Successful alternative strategies might include reliance on market incentives (for instance, consumer choice options [discussed below]), professional education, private accreditation, other privately sponsored quality assurance initiatives, and innovative payment systems that provide positive incentives to motivate desired kinds of behavior. Achieving the best combination of legal and private strategies requires accurately assessing the workings of each component, including legal regulation. Regulation is worthwhile only when it works more effectively and efficiently than other feasible options. Put differently, a wise utilization of limited resources compels asking whether the benefits of pursuing a particular regulatory strategy outweight the opportunity costs of foregoing other avenues to achieve the same desirable ends, and what factors could be worked with to change the benefits/costs ratio.

These two rationales for engaging in a TJ exercise ultimate reduce to the ethical principle of nonmaleficence, or *primum non nocere* (First, do no harm). This precept, which is just as valid in the legal as in the medical context where the principle is more usually discussed, means that we have an obligation not to make situations worse than they would be otherwise. Laws that have a negative or counterproductive effect on their intended beneficiaries violate this principle.

3.2 Application of the TJ Analytical Lens

3.2.1 Guardianship

Every jurisdiction in the United States has enacted statutes empowering the courts to appoint a surrogate decision maker with official authority to make decisions on behalf of a mentally incompetent ward. The terminology for the court-appointed surrogate decision maker varies among jurisdictions; "guardian" is the most commonly used term, although "conservator" and other terms are used in some locations.

In enacting guardianship laws, the state exercises its *parens patriae* power to protect those persons who cannot take care of themselves at a level that society believes is minimally sufficient, even if that means protecting people from the folly of their own decisions and over their own objections. The contemporary ethical and policy rationale for imposing guardianship on an unconsenting adult, therefore, is one of therapeutic benefit for that person, i.e., guardianship will be in that person's best interests.

For more than two decades, there has been a strong national movement in the United States to "reform" state guardianship laws. The focus has been on strengthening the substantive and procedural protections available to prospective wards, or "allegedly incapacitated persons" (AIPs). Reformers have advocated the enactment of enhanced due process requirements pertaining to, among other things: (a) the use of simple but specific guardianship petition forms; (b) the AIP's right to timely notice of the guardianship petition; (c) a mandatory right to legal counsel and a fair hearing, including the AIP being present at the hearing, compelling the attendance of witnesses, presenting evidence, and confronting and cross-examining all witnesses; (d) a minimum standard of proof of clear and convincing evidence; and (e) the right to appeal any adverse orders or judgments.

In response to these reform efforts, every state has modified its guardianship statutes, often very extensively, in both substantive and procedural respects. In terms of strengthening procedural safeguards, many states have enacted versions of the recommended reforms listed in the previous paragraph. Regarding substantive due process, various states have: changed their statutory language pertaining to incapacity in an attempt to reduce stigma; modified their definitions of decisional incapacity to emphasize function rather than old age; authorized and encouraged their courts to consider appointing guardians with strictly limited powers corresponding precisely to the ward's specific mental deficits; and imposed specific methodological requirements pertaining to the professional assessment of the ward's decisional capacity and the contents of the assessment report.

In light of all this legislative activity, the TJ question to be asked is: How well does the modern guardianship system "work"? This is a difficult query, in large part because of a continuing deep ambiguity, even schizophrenia, regarding the proper role of the guardianship system. For example, should a fundamental goal of a therapeutic guardianship system be to foster the creation of many new guardianships, because guardianship is primarily a form of therapy that compassionately, benevolently protects vulnerable persons against neglect and exploitation? Or, conversely, should success be measured by how few new guardianships are established, because guardianship should be viewed mainly as a limiter of individual rights, something to be invoked only as a last resort because too often this legal tool is misused as a dangerous excuse for turning state benevolence into excessive paternalism?

There exists a huge conceptual chasm between the therapeutic versus adversarial models of guardianship. In the former, the attorney is a helping, therapeutic agent guided by the professionally perceived needs of the proposed ward and working in concert with the other helping participants in the system. In the latter, the attorney functions instead as a zealous advocate for whatever choices the client expresses, protecting the rights of the client from the system and its attempted intrusions.

Evidence for judging the success of the guardianship system under either the therapeutic or adversarial models is rather limited. The evidence that has been compiled supports mixed conclusions. One problem in evaluating the TJ implications of current guardianship law is that statutory provisions can only be effective if they are properly implemented and enforced by the local court, law enforcement, and social service bureaucracies responsible for running the system and dependent on legislative appropriations for that purpose. Frequently, the intent of the laws does not get carried out correctly, especially when inter-agency collaboration is lacking and/or legislatures are unwilling to fund compliance.[1]

Another shortcoming of the evaluation record is that studies attempting to measure the impact of statutory modifications on guardianship practices or processes (e.g., diversion to less restrictive alternatives, minimizing use of emergency procedures, presence of the AIP at the hearing, use of functional rather than categorical evaluation, and use of limited instead of plenary (total) orders) (see e.g., Moye et al. 2007 p. 425) do not necessarily give us much useful information about the laws' impact on tangible results. It is, of course, the latter matter – the concrete outcome of legal interventions on the lives of identifiable intended beneficiaries – that is at the crux of a TJ inquiry. As explained by one legal commentator:

> The application of TJ analysis to adult protective proceedings is a natural step. Since the justification for such proceedings is the protection of the well-being of the protected person under the *parens patriae* power of the state, it is logical to judge different statutory schemes

---

[1] US Gov't Accountability Off., Guardianships: Little Progress in Ensuring Protection for Incapacitated Elderly People, GAO-06-1086T (2006).

> by evaluating how protected persons actually fare when placed under protective proceedings. The central question is whether the well-being of protected persons is improved compared to that of similarly impaired individuals who have not been the subject of protective proceedings (Wright 2004, p. 53, 73).

Laudable guardianship processes ought to, but do not always, produce desirable results for the real human beings upon whom those processes work. Under a TJ orientation, it is the probability of achieving particular desired outcomes that should dictate the processes best suited to achieving those outcomes. "In order to maximize the therapeutic possibilities of adult protective proceedings, we need to be prepared to discard old structures and procedures which have failed to achieve their purpose and to creatively design new structures and procedures to achieve the specific 'real world' outcomes that we desire" (Wright 2004, p. 83)

3.2.2 Nursing Home Regulation and Litigation

Some older individuals, because of a combination of severe impairments in the ability to conduct Activities of Daily Living (ADLs) and a less than ideal family/social situation, have no alternative to nursing home care. In the United States, the predominantly proprietary nursing home industry is characterized by pervasive, comprehensive, directive governmental regulation as the chief social strategy for assuring quality of care and quality of life for residents (most of whom are well over the age of 65). This multifaceted regulatory atmosphere is comprised, first, of the federal Nursing Home Quality Reform Act of 1987, as implemented by regulations promulgated by the Department of Health and Human Services that nursing homes wishing to participate in the Medicare and Medicaid programs are obliged to obey. Additional regulation is provided by state licensure acts, the Americans with Disabilities Act, the Rehabilitation Act, various criminal statutes relating to abuse and neglect, criminal and civil statutes outlawing health care fraud, and civil malpractice lawsuits alleging tortious conduct and/or breach of contract.

A TJ approach concentrates on the impact of nursing home regulations on the lives of residents, not on ordering providers around for its own sake. As one geriatrician has observed:

> In an area like long-term care, where so little is established about the relationship between process and outcomes, there is a strong argument for concentrating regulatory activities on assuring that satisfactory outcomes are achieved. Such a philosophy is at odds with practice. Often when uncertainty about the best path to follow is greatest, the press for orthodoxy becomes most intense (Kane 1998, p. 232).

The concept of TJ seeks to upset that orthodoxy.

How, then, does nursing home regulation in the United States measure up under a TJ analysis? The verdict, thus far, is mixed.

On the positive side, for example, nursing homes are found to perform the overwhelming majority of required resident assessments on time and to submit resident records to their state Minimum Data Set (MDS) databases in a timely fashion.[2] (This, of course, is a process rather than an outcome measure.) At the same time, credible academic, government, and consumer advocate critics of the current quality of nursing home care – now more than two decades since Congress massively redesigned the federal regulatory landscape for nursing homes – abound and persist in pointing out the many serious deficiencies that jeopardize resident care despite the pervasive regulatory environment.[3] Even when there is technical compliance with regulatory process requirements, actual improvements in the residents' quality of care and quality of life may not automatically be forthcoming.

The fear of malpractice suits brought by or on behalf of residents against the facilities and their staff – an apprehension that did not widely exist even a decade ago – has a particular ability to negatively affect quality of care by, among other things, making it more difficult for facilities to hire and retain adequate numbers of good staff. Access to needed services is affected negatively when providers' anxieties about potential litigation and liability, sometimes coupled with liability insurance unavailability or unaffordability, incentivizes providers to exit the nursing home industry, shrink in size, reject applicants (or families) they fear may be litigious, or operate without liability insurance.[4] The proliferation of civil claims being filed based on allegations of substandard care does not seem to have reduced (let alone eliminated) perceived instances and patterns of objectionable behavior by nursing homes. As one critic notes, "These examples of reckless abandonment of care [previously described in the cited article] leave numerous questions about why and how this could have happened" (Schall 2006, pp. 151, 152)

Moreover, litigation (or to a large extent even just its specter) drains needed cash from nursing homes in the form of insurance premiums, attorney fees, and insurance deductibles, in effect creating a sort of death spiral; *viz.* deficient facilities have less money to pay for needed staff, so instances of neglect increase, thereby creating more damage awards, which further drain care, and so on and so forth. The TJ implication is that private pay non-plaintiff residents pay more and all residents get less care.

---

[2] Off. Inspector Gen., US Dep't Health & Hum. Serv., Nursing Facility Performance in Assessing Residents Timely and Submitting Required Minimum Data Set Records, OEI-06-02-00730 (2005).

[3] US Gov't Accountability Off., Nursing Home Reform: Continued Attention is Needed to Improve Quality of Care in Small but Significant Share of Homes, GAO-07-794T (2007); US Gov't Accountability Off., Nursing Homes: Efforts to Strengthen Federal Enforcement Have Not Deterred Some Homes from Repeatedly Harming Residents, GAO-07-241 (2007); US Gov't Accountability Off., Nursing Homes: Despite Increased Oversight, Challenges Remain in Ensuring High-Quality Care and Resident Safety, GAO-06-117 (2005); *Nursing Homes – Business as Usual*, Consumer Rep'ts 38 (Sept. 2006); Off. Inspector Gen., US Dep't of Health & Human Serv., Preadmission Screening and Resident Review for Younger Nursing Facility Residents with Serious Mental Illness, OEI-05-05-00220 (2007).

[4] Medical Liability in Long Term Care: Is Escalating Litigation a Threat to Quality and Access? Hearing Before the Special Committee on Aging, US Senate, 108th Cong., 2d Sess., Serial No. 108–39 (July 15, 2004).

3.2.3 Home Health Care

An increasing number of older persons with impairments in the ability to perform ADLs are receiving long term care services in home and community-based settings rather than nursing homes. Home- and community-based long term care may consist of a number of different kinds of supportive services for the client provided by a variety of formal and informal caretakers, but most of the regulatory attention thus far has been focused on the operations of home health agencies (HHAs).

The regulation of HHAs and their staffs has been extensive and complex for many years. Individual states each require the mandatory licensure of both HHAs themselves and the various health care professionals who may be employed directly by, or work under a contractual relationship with, HHAs. Further, HHAs who wish to be certified to participate in the Medicare financing program also must satisfy the Conditions of Participation promulgated by the federal Centers for Medicare and Medicaid Services (CMS).

Because home health care regulation has mainly entailed an emphasis on structure and process measures rather than outcomes, research examining command-and-control regulation in this arena through anything resembling a TJ prism has been scant and inconclusive. There is little persuasive evidence that the agency licensure requirement or the Medicare standards and survey and certification process have accomplished their main objective of excluding deficient HHAs from participating in the Medicare program or doing business at all. Complaints about quality deficiencies persist, presumably with antitherapeutic implications for HHA clients, despite the pervasive regulatory environment.[5]

3.2.4 Consumer Choice and Direction

In the United States, government-financed health care, long term care, and retirement income programs designed exclusively or primarily to benefit older persons (respectively, the Medicare, Medicaid, and Social Security programs) have operated according to the model of command and control rules about eligibility, benefits, and procedures set up and enforced by the government. The paradigm has been evolving, however, toward one in which the individual beneficiary (or service "consumer"), or the surrogate acting on behalf of a decisionally incapacitated beneficiary, exercises a high degree of meaningful choice, direction, and control regarding the details of his or her publicly-funded benefits. These details are the mundane but essential factors that, in large part, really determine the quality of daily life enjoyed by older individuals who are both in need of health care, long term care, and financial services and dependent on public financial support to receive them.

[5]US Gen. Acct. Off., Medicare Home Health Agencies, Weaknesses in Federal and State Oversight Mask Potential Quality Issues, GAO-02-382 (July 2002).

3.2.4.1 Long Term Care

Traditionally, the federal and state governments directly regulated the what, where, who, when, and how details of long term care financed by the Medicare and Medicaid programs. Under this dominant funder- or agency-driven model, the public payer/insurer covered only eligible expenses, for specified amounts and kinds of services, delivered by approved providers. This model is predicated on a fundamental distrust by gerontological professionals and public policy makers of the ability of older persons who need services and support provided by others to safely manage their own affairs.

Today, though, the ageist mind set that has treated all older persons homogeneously, as frail, pathetic victims rather than as autonomous agents, is being challenged. Models of service delivery and financing that embrace proactive consumer choice and responsibility regarding important long term care details (such as hiring, paying, scheduling, supervising, and/or terminating one's own home care workers) are increasingly popular in the United States and elsewhere. A significant consensus has formed around the idea that intrusive regulation often acts less as an effective accountability mechanism than as a barrier to enabling desired and desirable consumer choice and professional discretion, making programs less—rather than more—responsive to their intended beneficiaries.

The key TJ question is how well consumer direction of long term care serves the consumer and any informal and formal caregivers. This is one area where a substantial body of empirical investigation has taken place and points overwhelmingly in one direction, namely, toward the conclusion that laws and policies introducing elements of consumer direction into publicly-financed long term care programs strongly serves the TJ goals of individual empowerment, enhanced quality of life, more flexibility and responsiveness to consumer needs by the service marketplace, and greater consumer satisfaction with the services received (see e.g., Symposium 2007 p. 353). At the same time, intrusive regulation makes fewer long term care options available to older consumers because, by raising the cost of compliance, it can serve as a barrier to entry for competing service providers.

3.2.4.2 Health Care

The Medicare program was enacted by Congress in 1965 to ease the financial burden on older and disabled Americans occasioned by their often expensive, extensive health care needs. The original legislation created two separate and distinct public health insurance programs for the elderly and disabled, traditional Medicare Parts A (mainly hospital coverage and some limited coverage for long term care) and B (paying chiefly for physician care, various outpatient services, and – under very limited circumstances – home health care). Under Parts A and B, the details of provider compensation – and therefore largely the particulars of health care delivery—were firmly established by federal statute and its implementing regulations and administrative guidelines.

As a sharp contrast to the command and control regulatory approach embodied in Parts A and B, Congress enacted Medicare Part C as part of the Balanced Budget Act of 1997 to create the Medicare + Choice Program. Part C provided an array of private fee-for-service and managed health insurance options for selection by Medicare beneficiaries. In the Medicare Modernization Act (MMA) of 2003, Medicare + Choice was converted into the new Medicare Advantage program, which offered even more health insurance options for beneficiaries. Each eligible older and disabled person now has the right to individually select between remaining in Parts A and B (i.e., traditional third-party insurance coverage with details determined according to federal regulation) or enrolling in any of the market-oriented options that are offered in the beneficiary's geographic region. The MMA also created a new Medicare Part D, which allows beneficiaries enrolled in any of the other parts of Medicare to also select a federally approved private prescription drug plan.

Critics of the sort of consumer choice and direction represented by Medicare Parts C and D think this approach is antitherapeutic. They assert that too many older persons are, and inevitably will be, too frail, weak, dependent, institutionalized, or otherwise vulnerable to exploitation to survive successfully in a less regulated, more marketplace characterized environment. According to this view, a market atmosphere in which prudent or lucky choices lead to "winners" but unwise or unlucky choices may create "losers" presents too many risks of danger to exposed persons, and those risks, if they materialize, are too severe for a compassionate society to tolerate. This is an ideology that defines therapeutic result in terms of risk avoidance, and conceives of government regulation as the ultimate risk manager for life's contingencies.

In contrast to this critical perspective, defenses of enhanced consumer choice and direction as a welcome replacement for extensive command and control regulation stem mainly from a commitment to the therapeutic possibilities of personal autonomy. The autonomy model works on the premise that respecting older individuals enough to allow and encourage them to make their own voluntary and knowing choices about the terms of their publicly-funded health care plans will be therapeutic for those persons. There is a growing body of research seeming to support this position (see e.g., Serota 2007, p. A16 (citing studies showing a high degree of satisfaction among Medicare Advantage enrollees)).

Of particular note, agency-driven models do not respond well to changes in the state of medicine. Over time, this deficiency results in a package of benefits that becomes outdated. For instance, Medicare covers the first 60 days in a "spell of illness" for a hospital stay, but today nobody – not even older persons – stays in a hospital that long. TJ is disserved when Medicare provides benefits that older people no longer need, rather than benefits (such as custodial long term care) that are much more important today than they were in 1965.

In addition, Medicare Part C offers the managed care opportunity for better coordination of care, compared to the Medicare Part A model of focusing on discrete episodes of acute illness. Care coordination is especially important and beneficial to older persons, the majority of whom have one or more chronic conditions.

3.2.4.3 Retirement Income

The main public source of retirement income in the United States is the Social Security Old Age program. This program has always been structured on a mandatory, defined benefit basis, in the sense that the federal government guarantees each conscripted beneficiary a definite amount of monthly income computed according to a complicated formula. To obtain this guaranteed financial benefit, the beneficiary has no choice but to contribute a percentage of income during the working years to the Social Security Administration's Old Age Trust Fund, whose assets may be invested by the federal government only in government financial vehicles according to narrow federal rules.

The Bush administration has issued several proposals for reform of the Social Security retirement program. These various proposals involve the possibility of more or less beneficiary control over the investment of the beneficiary's own Social Security retirement funds, including proposals that would allow private investment of at least a portion of those funds. Fundamental to these proposals is the idea that the Social Security retirement income program ought to operate more as a defined contribution model in which the investor/beneficiary enjoys the possibility of achieving greater returns through the personal investment of retirement funds in private securities than would be assured under the prevailing defined benefit program.

The potential therapeutic or antitherapeutic impact of a more defined contribution approach to public retirement income funds should be defined broadly, taking into account psychological factors (the self-esteem and satisfaction of autonomous control versus the anxiety caused by risk of financial loss) as well as the obvious dollars and cents ramifications. At this time, estimating the possible magnitude of any of these factors would be an exercise in almost pure speculation, but research done on carefully designed small pilot demonstrations of the defined contribution plan approach could provide useful data to inform a TJ analysis in this area. Any changes in the Social Security statute and regulations could then be tailored around the evidence produced.

3.2.5 Research Participation

The scientific frontiers making available better, more therapeutic treatment of older persons with severe mental disorders such as dementia and depression can be pushed forward only by encouraging more biomedical, behavioral, and health services research concerning individuals with such disabilities. The legal and ethical Catch 22, however, is that useful research in this arena must often involve the participation as subjects of individuals who – precisely because of their illnesses – lack adequate cognitive and emotional capacity to engage in a rational decision making process leading to autonomous consent or refusal for research participation.

Although society, as well as present and future mentally impaired people, have a valid interest in promoting, even facilitating, the conduct of relevant research, there is also a social responsibility to protect persons with severely impaired decisional capacity against abuse or exploitation in the research context. One potential regulatory response to this conundrum would be to impose an outright ban on the enrollment of seriously decisionally impaired individuals in any research study. This Draconian response probably would be sustainable as a legitimate exercise of Congress' constitutional authority to regulate interstate and foreign commerce or the states' inherent *parens patriae* power to protect people who cannot adequately protect themselves. However, an outright ban would be morally objectionable for several reasons.

First, a prohibition would essentially eliminate any meaningful chance for making additional scientific or programmatic progress in diagnosing, preventing, curing, or caring for exactly that group of persons supposedly being shielded from harm, namely, those who are afflicted with dementia, depression, psychoses, and other disorders jeopardizing decisional capacity. Eliminating the opportunity for such progress risks creating a class of "therapeutic orphans" (Backlar 1998 p. 829). It is not surprising, therefore, that advocacy organizations for individuals with such disorders and their families are among the biggest boosters of including the afflicted as research subjects.

Second, "[P]rohibiting such research might harm the class of mentally infirm persons as a whole by depriving them of benefits they could have received if the research had proceeded."[6] Those benefits are not just limited to the scientific advances in diagnosis or treatment that might result from the research; the concern here is also with psychological benefits being denied to individuals who would be legally prohibited from participating in research that would have studied precisely their own situation. The sharp insight of one set of commentators is reflected in the comment:

> Respecting self-determination or autonomy, even when reduced, can have important therapeutic value. An individual's sense of dignity and personhood would be frustrated when his choices are rejected on the ground that he is incompetent. The literature on the psychology of procedural justice stresses the importance of voice and validation, and of being treated with dignity and respect.…For the same reasons, people seeking to participate in research that they think will be beneficial for them will feel that they are being treated without dignity and respect when they are prohibited from such participation on the basis that their choice is not sufficiently competent (Winick and Goodman 2006, pp. 485, 489–490).

These commentators further observe that the antitherapeutic effects of labeling the individual as incompetent to decide about research participation might include negative self-attribution, diminishing functioning and motivation, and producing a form of learned helplessness and clinical depression. For the families and support

[6] National Commission for the Protection of Human Subjects of Biomedical and Behavioral Research, Report and Recommendations: Research Involving Those Institutionalized as Mentally Infirm 58 (1978).

groups of individuals with diminished mental capacity, thwarting research because normal informed consent is impossible may be harmful in terms of the sense of frustration, powerlessness, helplessness, and depression generated by their ability to do nothing to alleviate their loved one's plight.

Lawmakers need to reevaluate how the informed consent doctrine ought to apply to this type of situation. Considerations of the potential therapeutic and antitherapeutic effects of different alternatives ought to be taken into account in choosing what deviations, if any, from strict adherence to the usual rules of informed consent ought to be allowed and even facilitated.

3.2.6 End-of-Life Medical Care

Various jurisdictions have undertaken a constellation of laws intended to improve the process of decision making by and for medical patients in their last stages of life, and thereby to achieve the therapeutic effect of enhancing the quality and dignity of the dying process. Among other things, every state has enacted statutes recognizing mechanisms for presently mentally capable adults to execute written advance instruction (e.g., living will) and proxy (e.g., durable power of attorney) directives as a way to maintain a degree of prospective autonomy for the future (see Chapter 2). Most states also have passed legislation spelling out the rights and responsibilities of the various pertinent parties in the event that no advance directive was timely executed but the patient is presently unable to make decisions. Additionally, a few hundred judicial decisions have tried to delineate parameters for end-of-life medical decision making, either by interpreting specific statutes and regulations or on the basis of constitutional or common law principles.

The tangible therapeutic results of these legal initiatives regarding end-of-life care have been, at best, mixed. The general consensus is that the law has had a very limited impact on improving the quality of medical care for dying patients, and that serious deficiencies in the humaneness of care for the dying in health care institutions persists despite the pervasiveness of a substantial legal presence. Indeed, it may well be that the substantial legal presence–both real and perceived, and sometimes quite exaggerated–itself often creates a powerful aura of anxiety among health care professionals and providers about adverse civil or criminal consequences that produces counterproductive, antitherapeutic patient care behavior driven more by risk management considerations than compassion and concern for dying patients and their families (Kapp 2002 p. 586, 2003 p. 111).

3.2.7 Miscellaneous Areas

There are many other areas where the law—for better or worse—affects exclusively or primarily older persons. Those areas most ripe for future TJ analyses of effectiveness for intended beneficiaries include: regulations prohibiting age-based discrimination in employment, housing, insurance, and other matters; mandatory and voluntary elder mistreatment reporting provisions and interventions; forensic investigations of testamentary capacity (Marson et al. 2004, pp. 71, 72–73 (citing Drogin 2000)); and rules affording persons' preferred status as parties in the civil or criminal justice systems (e.g., expedited placement on the court's docket, leniency in sentencing) based solely on chronological age. Research needs to be conducted to shed light on whether these well-intentioned laws actually improve the quality of elders' lives as intended or perversely reinforce ageist stereotypes and thereby harmfully stigmatize members of the group intended to be helped.

3.3 Future Direction

The law, as one important instrument of public policy, has the potential to exert an enormous influence – for better (i.e., therapeutically) or worse (i.e., antitherapeutically) – on the real, lived lives of older individuals. I have opined previously: "Public policy making for the elderly ought to be a continuous, iterative process for which improvement in content depends (or ought to depend) on accurate feedback in response to these [TJ] kinds of inquiry" (Kapp 1996, pp. 3–4). This will happen only if the public policy debate and resultant lawmaking is thoroughly informed about the true therapeutic or antitherapeutic impact of specific laws and legal practices on intended older beneficiaries, and such information will only emerge if much more attention is devoted to the systematic, empirical study of these matters.

Besides legal and social science scholars carrying out (and funders supporting) more rigorous research initiatives aimed at filling in the TJ knowledge gap regarding the effects of law and legal practice on older persons' quality of life, it is essential that elder law practitioners be sensitized to the values and goals of TJ so they can incorporate into their everyday legal practices attitudes and techniques calculated to mitigate the antitherapeutic effects, and enhance the therapeutic impact, of involvement with the legal system for older clients.[7] Client-centered professional practice sensitive to TJ considerations will provide legal assistance of excellent technical quality while simultaneously focusing on the older client's psychological well-being.

The law is not an end in itself; rather, it is a necessary instrument to move societies toward desired practical goals. TJ is a valuable tool for telling us how well or poorly the law is doing in moving us toward the goal of improving the quality of life for older persons.

[7] Regarding the use of clinical legal education to shape elder law practice in a direction that maximizes the therapeutic effect for the older client, see Faulkner 2005, p. 685, 687 (citing Stolle et al 1997); Wright 2005.

References

Backlar P (1998) Anticipatory planning for research participants with psychotic disorders like schizophrenia. Psychol Pub Pol & L 4:829

Doron I, Fal I (2006) The emergence of legal prevention in old age: findings from an Israeli exploratory study. J Cross Cult Gerontol 21:41

Drogin EY (2000) From therapeutic jurisprudence…to jurisprudent therapy. Behav Sci & L 18:489

Faulkner CE (2005) Therapeutic jurisprudence and preventative law in the Thomas M. Cooley Sixty Plus, Inc., Elder Law Clinic. St. Thomas L Rev 17:685

Kane RL (1998) Assuring quality in nursing home care. J Am Geriatr Soc 46:232

Kapp MB (1996) Therapeutic jurisprudence and older lives: Well-intended laws and unexamined results. J Ethics, L & Aging 2:3

Kapp MB (2002) Regulating the foregoing of artificial nutrition and hydration: First, do some harm. J Am Geriatr Soc 50:586

Kapp MB (2003) Legal anxieties and end-of-life care in nursing homes. Issues in L & Med 19:111

Marson DC, Huthwaite JS, Hebert K (2004) Testamentary capacity and undue influence in the elderly: A jurisprudent therapy perspective. L. & Psychol Rev 28:71

Moye J et al (2007) Statutory reform is associated with improved court practice: Results of a tri-state comparison. Behav Sci & L 25: 425

Schall VA (2006) The new extreme makeover: The medical malpractice crisis, noneconomic damages, the elderly, and the courts. Appalachian J L 5:151

Serota SP (2007) Advantage, Seniors, Wall St. J., April 2, p. A16

Stolle DP, Wexler DB (1997) Therapeutic jurisprudence and preventative law: a combined concentration to invigorate the everyday practice of law. Ariz. L. Rev. 39:25

Stolle DP et al (1997) Integrating preventative law and therapeutic jurisprudence: a law and psychology based approach to lawyering. Cal W L Rev 34:15

Stolle DP, Wexler DB, Winick BJ (eds.) (2000) Practicing therapeutic jurisprudence: Law as a helping profession

Symposium (2004) Advocacy and aging, Generations 5

Symposium (2007) Putting consumers first in long-term care: findings from the cash & counseling demonstration and evaluation. Health Serv Res 42:353

Wexler DB, Winick BJ (eds.) (1991) Essays in therapeutic jurisprudence

Wexler DB, Winick BJ (eds.) (1996) Law in a therapeutic key: developments in therapeutic jurisprudence

Winick BJ, Goodman KW (2006) A therapeutic jurisprudence perspective on participation in research by subjects with reduced capacity to consent. Behav Sci & L 24:485

Wright JL (2004) Protecting who from what, and why, and how? A proposal for an integrative approach to adult protective proceedings. Elder L J 12:53

Wright JL (2005) Therapeutic Jurisprudence in an Interprofessional Practice at the University of St. Thomas Interprofessional Center for Counseling and Legal Services. St. Thomas L. Rev. 17:501

Chapter 4
A Feminist Approach to Elder Law

A.K. Dayton

Global aging is a phenomenon of which academics in a broad range of disciplines appear to be acutely aware. In recent years, the popular media and many reputable scholars have seized upon the metaphor of population aging as the "grey dawn"[1] or "impending earthquake" or "tsunami" that threatens the global economy.[2] This apocalyptic vision of current demographic trends is problematic, because it invites a perception of the world's elderly not as an important resource to be valued and cherished, but as a drain on "the rest of us" and a target of fear and loathing. Such a vision cultivates and legitimizes pervasive ageism and diverts attention from useful discussions about how public resources ought to be used to ensure a quality of life for all the world's citizens, not just those who are perceived to be current or future contributors to the tax base that sustains national economies.

Apocalyptic demography is of particular danger to women due to the feminization of aging (see generally Cruikshank 2003). One important truth about global aging is that it is a "women's issue". Women live longer than men in all but a handful of countries, developed and developing alike. In the world's oldest nation, Japan, the ratio of men to women at age 65 is 0.732[3]; 85% of that nation's centenarians are

---

[1] E.g, Peterson (1999):
[G]lobal aging, like a massive iceberg, looms ahead in the future of the largest and most affluent economies of the world. Visible above the waterline are the unprecedented growth in the number of elderly and the unprecedented decline in the number of youth over the next several decades. Lurking beneath the waves, and not yet widely understood, are the wrenching economic and social costs that will accompany this demographic transformation-costs that threaten to bankrupt even the greatest of powers, the United States included, unless they take action in time.

[2] E.g. Hiemstra (2005):
[A] real tidal wave is approaching. Everyone has heard of it, so I suppose many think, since it is old news, it must have been addressed. It is the tidal wave of an aging population. It turns out that a tidal wave, or tsunami, may be an apt analogy. How did the recent tsunami manifest? Not in a single dramatic wave, but rather a relentless series of surges that overwhelmed what lay in their path."

[3] US Central Intelligence Agency, World Fact Book, Field Listing – Sex ratio, https://www.cia.gov/library /publications/the-world-factbook/fields/2018.html

I. Doron (ed.) *Theories on Law and Ageing: The Jurisprudence of Elder Law*,

women.[4] In China, the country with the largest number of elderly persons, elderly women outnumber men by some five million.[5] Elderly women are more likely to live in poverty and to live alone or be relegated to old-age homes that provide marginal or substandard care. Yet, virtually without exception, laws and social policies affecting such matters as entitlement to public pensions, the delivery and financing of long term care, and the allocation of public resources to prevent victimization and abuse of vulnerable adult, are infused with and reflect with historic patterns of overt and covert discrimination against women in the workplace and the political arena, and perpetuate the systematic devaluation of "women's work." As a consequence, as nations age, the convergence of demographics and imbedded patterns of discrimination against women throughout their lives with respect to health care, employment, and the demands of the private sphere results in an aging population that suffers disproportionately from poverty, isolation, and abuse. Laws and policies that fail to provide for the needs of the elderly population create enormous financial burdens on younger women who are the primary caregivers of older relatives.[6] It is thus imperative that scholars examine and expose the legal framework defining the personal, health, and income security of their oldest citizens in light of the reality that most elderly persons, as well as those who care for them, are women. Reform of elder law and policy must take account of this reality and assure that such "reform" does not exacerbate existing discrimination against and injustice towards women.

The extent to which lawmakers understand the factors that contribute to the phenomenon of the feminization of aging, and the depth of their commitment to address its consequences, is far from evident. Feminist jurisprudence as a vehicle for explaining and reforming elder law and policy has been largely ignored by academics, policymakers, and politicians alike. Accordingly, this chapter suggests how current law and aging policy generally fails to account for the lives of women and suggests how a feminist approach to elder law would ultimately enhance the quality of life for all elders, irrespective of their gender, race, or family status.

4.1 Principles of Feminist Theory in Law

Feminist jurisprudence began to make an impression in the legal academy somewhat after second-wave feminism had begun to pervade other disciplines. Articles discussing "feminist jurisprudence" did not appear in law reviews until the early 1980s; the first anthology of feminist legal scholarship was not published until 1991 (Bartlett and Kennedy 1991). In the past twenty-five years, the term "feminist jurisprudence" has become an umbrella term for a broad range of perspectives on

[4] US Census Bureau, International Data Base, Country Summary: Japan, http://www.census.gov/ipc/ www/idb/country/japortal.html

[5] GeoHive, Population, http://www.xist.org/charts/pop_agestruc.aspx

[6] See generally Family Caregiver Alliance, Women and Caregiving: Facts and Figures, http://www.caregiver.org/caregiver/jsp/content_node.jsp?nodeid = 892

"women and the law" that incorporate not merely gender, but race, ethnicity, class, sexual orientation, and other attributes of self. The goal of this chapter is not to offer a comprehensive discussion of feminist jurisprudence, but to summarize the primary components of modern feminist legal theory and illustrate how a feminist approach to law and social policy affecting the elderly differs in scope and objective from other theoretical approaches. Accordingly, the overview of feminist jurisprudence that follows is brief, and is intended to paint with a broad brush.

Liberal feminism. Liberal feminism, sometimes called equality feminism, is in a sense the least controversial branch of feminist theory. Its essence can be a captured in bumper stickers that read "A woman's place is in the House…and in the Senate", and is represented by the writings of nineteenth century feminist Mary Wollstonecraft as well as the early scholarship of Wendy Williams (e.g., Williams 1982). Liberal feminist jurisprudence suggests that full equality for women can be achieved simply by "evening the playing field" – that is, giving women access to the political process, to educational opportunities, and to the workplace, primarily by outlawing discriminatory practices that deny women rights that men are assumed to or do have. Laws that are either facially gender-neutral, or that bar discrimination based on gender, will generally satisfy the demands of the liberal feminist purist. The 19th amendment to the United States Constitution, which gave women the right to vote in national elections, is a manifestation of liberal feminist principles, as is Article 23 of the European Union's Charter of Fundamental Rights,[7] and Title VII of the US federal Civil Rights Act of 1964 (prohibiting employment discrimination against women).[8] A liberal feminist jurisprudence demands that women be treated "in the same way" that men are treated; it barely looks beyond this relatively uncontroversial account of "equality" to examine the social and cultural underpinnings of the positive law or the institutional traditions that the law affects and fosters.

Cultural (difference) feminism. As does liberal feminism, cultural feminism embodies relatively straightforward principles. The most critical limitation of liberal feminism is that it suggests that women should have the rights that men have "because they are men" only to the extent that women are like, or become like, men. But there are biological differences between women and men – most notably, the ability of human females to bear children and the relative physical prowess of men – that make "equal treatment" of them in the marketplace impossible. Modern cultural feminism draws heavily from the work of psychologist Carol Gilligan, who demonstrated that research which purported to identify the "norms" of the stages of moral development" was based solely on references to the moral development of boys (Gilligan 1982). When measured against these "norms", girls inevitably fail to achieve full "moral development" – their method of resolving ethical dilemmas rarely "progressed" to the absolute rights-based approach to which boys generally "evolve." Cultural feminism admits the possibility of empirically-verifiable differences between girls' and boys' (hence, women's and men's) approaches to

---

[7] http://eur-lex.europa.eu/LexUriServ/LexUriServ.do?uri = CELEX:32000X1218(01): EN:HTML

[8] 42 USC §§ 2000e et seq.

moral reasoning and decision-making but holds that, whether these are based in biological or culturally-induced gender differences, the male approach should not be privileged or held out as a universal norm. According to Gilligan, women's "voices" respecting moral reasoning are different from, but are not inferior to, those of men. On this view, women's greatest contribution to moral and political discourse stems from the perspectives they offer from their roles as child-bearers, caregivers, and mediators.

Cultural feminist jurisprudence contends that biological and culturally-constructed differences between women and men will inevitably preclude the ability of a pure liberal feminist approach to correct inequality, and calls for accommodation of these differences in both public and private life. The failure of liberal feminism to account for, at a minimum, the inherent biological differences was starkly captured *Geduldig v. Aiello*,[9] in which the United States Supreme Court held that a state disability compensation scheme that covered virtually all disabilities except those relating to pregnancy was not discriminatory because it did not distinguish between men and women, but rather "pregnant women and nonpregnant persons."

Modern cultural feminism embraces the 19th century construct of "separate spheres" – the domestic sphere of family within which women predominately operate, and the business sphere within which men operate – but claims that the work women do within the domestic sphere should be valued to the same extent as the work men do out side it. Thus, women who become pregnant cannot be fired due to pregnancy; instead, the importance of child-bearing should be formally acknowledged and accommodated through legal protections for the condition of pregnancy that reflect the value societies purportedly place on child-bearing. Ultimately, cultural feminist legal theory found its way into laws and workplace policies that treat pregnancy as an illness (for purposes of accounting for sick leave), maternity and eventually paternity leave policies, and caregiver leave statutes that mandate employers to provide unpaid time off work to provide care for an ailing family member. Some of cultural feminism's most visionary objectives – implementation of comparable worth doctrines in the workplace, for example – were not accomplished, but in some respects the cultural account of feminist theory should be credited with effecting significant positive change in the direction of equality in some areas of the public sphere.

Radical feminism. Radical, or dominance, feminism is certainly the most caricatured of all variations of feminist thought.[10] Its roots lie partly in the writings of 19th century abolitionist Sarah Grimké (1837), and it is often linked with Marxist thought as well (MacKinnon 1983). The most basic premise of radical feminism is that the evolution of women's inequality in the public and private

[9] 417 US 484 (1974).

[10] US Presidential wannabe Pat Robertson, for example, claimed at the 1992 Republican National Convention that "Feminism is a socialist, anti-family, political movement that encourages women to leave their husbands, kill their children, practice witchcraft, destroy capitalism and become lesbians."

spheres is tied to the physical power that men have historically had over them (hence, "dominance") and the corresponding threat of violence, including sexual violence, that this power connotes. A radical feminist approach to law questions whether the "different voice" touted by the cultural feminist camp is really biologically based. Recalling Sarah Grimké's famous metaphor, radical feminist Catherine MacKinnon in 1987 wrote, "Take your foot off our necks, then we will hear in what tongue women speak (see e.g., MacKinnon 1987, p. 45)."

Radical feminism claims that the law and all other cultural, social, and political institutions perpetuate male dominance by simultaneously and paradoxically ignoring its existence and privileging it. It contends that both equality theory and difference theory accept the concept of a male "norm" and attempt to work around that norm but not challenge it (see e.g., MacKinnon 1987, p. 45). The claim that partriarchy – men's favored position at home and in the workplace – is achieved through direct or covert coercion has important implications. To accomplish true "equality" for women, radical feminism asserts, male privilege must be eradicated from cultural, social, and political institutions. Only if these institutions are completely restructured to account for women, however they may be defined biologically and socially, can true gender equality be attained. Remedial actions that achieve this outcome are permissible, even if they allow women access to "rights" that are not available to men.

Radical feminist jurisprudence underlies the concept that women subjected to a barrage of sexual innuendo in the workplace are experiencing employment discrimination because this creates a hostile environment in which women are devalued and marginalized, and from which many women will flee. The theory that a hostile environment equals sexual harrassment equals employment discrimination, without some *quid pro quo*, no longer seems controversial, but it certainly did when the theory was first proposed by MacKinnon in 1979 (MacKinnon 1979). Radical feminist jurisprudence also holds that the value of achieving equality for women has the same status as the need to protect other civil liberties such as freedom of speech; thus, they contend, the free speech rights of pornographers are not absolute, and must be balanced against the empirical likelihood that certain kinds of pornography may incite or encourage violence against women and their general subjugation at home and work. Under this approach, pornography can be restricted or even banned; moreover, women ought to have civil remedies against those who promote violent, demeaning, or hateful images of women as a remedy for the harm it causes individual women and women as a class (MacKinnon 1995). The pragmatic goal of radical feminism is to effect legislative change in the direction of eliminating patriarchy and redistributing power to women in both the public and private spheres.

A note on essentialism, multicultural feminism, and related variations on a theme. A frequent criticism of much of the early foundational feminist scholarship is that it ignores or trivializes the unique perspectives of those who face discrimination and oppression not only because of their gender, but due to other aspects of self – skin color, ethnicity, sexual identity, class, disability, etc (see e.g., Harris 1990 p. 581). This "essentialist" critique suggests that the experiences of upper middle class white women such as MacKinnon are at least as different from those of a poor black woman as they are from those of men. Each descriptive characteristic that can be assigned to a particular woman affects both her experience in the world and her perspective of it; the farther one is from the wealthy white male center of the universe, the greater the degree of oppression one experiences. Indeed, the notion of "standpoint epistemology" can be understood to contend that the more remotely from that center one is situated, the better one's understanding of the true nature of oppression. It is ironic, therefore, that even so-called anti-essentialist feminist discussions usually overlook the role that aging plays in the oppression of all women inasmuch as all women (if they live long enough), will become old (see, for example, Dowd and Jacobs 2003 (in which not one of the 46 essays and excerpts included in the book addresses the convergence of gender and age discrimination)). One goal of this chapter is to remind feminist scholars that as women pass into the realm of the elderly, they will face new forms of discrimination and oppression deriving from the confluence of the lifelong gender discrimination that all women face, and age discrimination that is pervasive and often unnamed or unrecognized in the public and private spheres.

Feminist theory has struggled to articulate the concept that most women, irrespective of race, sexual preference, or income, face similar barriers in some aspects of social and political life, but also face discrimination and disenfranchisement that is qualitatively and quantitatively different from gender-based oppression. It is beyond the reach of this short chapter to do more than acknowledge that struggle and recognize that the discussion is intended not as a comprehensive account of law, aging, and feminism, but rather as a springboard for future discussion. In the pages that follow, the absence of explicit references to the roles of race, sexuality, and characteristics of particular women that ultimately affect the degree and nature of the domination and inequality that she is likely to experience should be taken only as a concession to space constraints, not a result of ignorance or disregard for the multiple sources of women's inequality and oppression.

4.2 Applications

Law and aging, or elder law, as an academic and practice specialty is relatively new, and its development in different nations is closely tied both to demographics and to national laws and policies pertaining to the financing of public pensions and health care. One can describe the field as encompassing a broad range of matters ranging from health care access to age- and disability based discrimination, elder abuse, consumer fraud aimed at the elderly, wills, trusts, and estates, surrogacy, guardianships and conservatorships, and mental health law. It is probably more useful to describe elder law as the particular manner in which any aspect of law touches the lives of older persons. Some laws and policies are developed particularly out of concern for the elderly – public pension systems are implemented to assure or at least contribute to the income security of the aged. Other legal institutions are age-neutral in their reach – universal health care, for example – but are of special significance to older persons due the greater need of the elderly for certain kinds of health care. In short, some areas of law have a greater significance to older persons and affect the elderly population more profoundly that other demographic groups. Because the elderly population is predominantly female, “elder law” and its reform are of particular significance to women.

The analyses set below focus two exemplary areas that should be considered significant components of elder law. It shows how laws and public policy affecting the aging would benefit from feminist reforms that would assure something closer to equal treatment under law for all elders irrespective of their gender.

4.3 The Financing and Delivery of Long-Term Care

Old age is often accompanied by declining physical and mental health that can result in a need for so-called “long term care.” This term is used to describe the amorphous combination of skilled nursing care, custodial/personal care, and (sometimes) room and board provided to persons who have disabilities that interfere with their capacity to perform the various activities – eating, bathing, shopping, housework, etc. – that are necessary to living independently. Long term care is usually provided by third party caregivers in an elder’s home, but in some countries, particularly the Scandinavian nations, there is a culture of moving the elderly to old age care homes or skilled nursing facilities once their abilities are significantly compromised. In those nations with a substantial population of older institutionalized persons, a very significant majority of that population is women.

As a general rule, most aspects of long-term care are not considered “health care” for purposes of universal health care programs such as Canada’s Medicare system or the UK’s National Health Service. Due to population aging in most of the world’s industrialized nations, the imminent need to devise effective means for delivering and financing long-term care for the elderly has become increasingly apparent. The general failure of health-care systems to address the long term care needs of an aging population is itself amenable to a feminist critique. As noted above, female family members (spouses, adult daughters and daughters-in-law) historically have been the primary providers of long term care to the elderly, in industrialized nations as well as in less developed countries. Although many men do contribute to family caregiving of aging parents and in-laws, women’s socially-constructed role as caregivers, so evident in the context of child-rearing, is largely replicated when parents age and require personal care or other assistance. Indeed, a great many women will spend more years providing care to their parents than to their children. The direct and indirect costs of this model of financing long term care fall heavily on women in the guise of reduced income due to time taken off from paid work to provide care, the resulting loss of potential future retirement-related benefits, and the physical and emotional toll that caregiving can take.[11]

[11] See generally Family Caregiver Alliance, 2007 National Policy Statement. http://www.caregiver.org/caregiver/jsp/content_node.jsp?nodeid = 1908

As the population of elderly persons in a society rises, birth rates fall, and more women seek employment outside the home, this model for serving the long term care needs of the senior population becomes problematic: there are not enough stay-at-home-caregivers (that is, adult daughters and daughters-in-law) to provide support for all those elderly who need assistance. In countries that have a tradition of institutionalizing significant numbers of elderly, occupancy rates within old-age care homes and skilled nursing home facilities may rise, resulting in bed shortages (as in Canada and Australia) or patient-dumping of the most physically and cognitively frail residents (as in the United States). Most developed nations have been compelled by the convergence of these demographic realities to look for new ways to deliver, and pay for, the expanded need for long term care. But to what extent have policymakers considered the problem of providing quality long-term care to the elderly from feminist perspective? The unfortunate answer is: almost never.

In a number of nations facing a long-term care "crisis", for example, one solution has been to look to filial obligation laws that have long been on the books but were rarely enforced. In the United States, for example, the state of Pennsylvania recently resurrected a long dormant filial obligation law by strengthening the law and defining more explicit financial obligations on the part of children to reimburse the state for any expenditures it may have made on long term care for their parents (see Pearson 2006). While the law does not distinguish between male and female children, and thus would appear to be "gender-neutral", the inevitable effect of such laws will be to discourage elderly persons from obtaining professional care – either in their homes or in a facility – due to the heavy financial burden this will impose on families. We know from existing data that women already assume a disproportionate share of responsibility for unpaid care; imposing financial responsibility on children does little more than shift additional responsibility to them, exacerbating these caregivers' long-term financial and personal vulnerability and, in many instances, compromising the quality of care provided to elders whose physical or cognitive impairments call for a level of care that cannot adequately be provided by non-professionals. In short, expanding "filial" responsibility for long term care of the elderly is not acceptable unless those women who will take on the bulk of the legal obligation to provide care are, at a minimum, accommodated in the workplace, as with tax credits, paid leaves, or other forms of compensation that will counterbalance the financial implications of what has traditionally been unpaid caregiving.

Consider also, from a feminist perspective, an "immigration solution" to an impending long term care crisis. Long-term care reform in some developed countries has entailed, in part, relaxing immigration restrictions to allow an influx of skilled and semi-skilled workers such as nurses and personal care attendants from the less developed world who will serve as low-cost caregivers of the aged. Italy, which has proportionately the world's second largest elderly population, has taken this approach with the result that hundreds of thousands of foreign workers now provide the bulk of caregiving to that nation's older disabled citizens (see generally, e.g., Screti 2005).

Using immigration law to increase the pool of low-paid family caregivers can be characterized not only as a form of modern colonialism, but as the antithesis to a feminist approach to long term care reform. Such a model exploits the low value that societies tend to place on "women's work" such as personal care, nursing, and similar activities that make up the range of services required by the frail elderly as they become more dependent. Home health care workers are often unprotected – either *de jure* or *de facto* – by minimum wage and hour laws.[12] An immigration model of long term care reform perpetuates an underclass of immigrant women who earn barely enough to support themselves but often send much of what they earn to families left behind in their native countries. In short, long term care reforms that focus on immigration policy are discriminatory and unjust, unless they are accompanied simultaneously by wage and hour reforms that assure immigrant caregivers a level of income that is commensurate with the important role they play in assuring an acceptable quality of life for all elders.

One example of a long-term care reform plan that arguably comes close to the feminist ideal is Japan's recently adopted national long term care insurance program (see Glickman 2007, no. 7–8). Japan is the world's oldest nation: almost 21% of its population is age 65 or older. Faced earlier than most nations with the demographic situation described above, Japan in 2000 implemented a long-term care payroll tax on all workers and retirees aged 40 and older. The revenue generated by this new tax supports an infrastructure of public and private long term care. Elderly persons and those younger individuals who have disabilities or impairments resulting from a condition that is normally associated with aging (e.g. Alzheimer's disease) are entitled to services paid by their long term care insurance based on their level of need. Those in need of assistance can choose how they wish to receive services – at home, in adult day care, at an institution, etc.

What is most interesting about the plan, from a feminist perspective, is that cash payments to beneficiaries are not allowed. Japanese feminists opposed early suggestions that benefits be distributed in the form of cash allowances on the ground that this would simply channel income to elderly households without relieving social pressures on daughters and daughters-in-law to become primary caregivers to the aged (Campbell and Ikegami 2000). Although this feminist insight has not created a perfect system – there is some evidence that a low-paid underclass of personal care attendants is developing in Japan – the inclusion of a feminist perspective in the policy discussions that led to Japan's current approach to financing and delivering care offers lessons for the rest of the world's nations as they seek solutions to the growing need to provide long term care to their oldest, most vulnerable citizens.

4.4 Social Security and Public Pension Systems

Most of the world's nations offer some kind of public old-age pension system.[13] There is no consistent terminology used to describe these systems regarding when

[12] See, e.g., Long Island Care at Home, Ltd. v. Coke, 551 US – (2007).

[13] For a thorough discussion of pension systems in 53 countries, see Edward Whitehouse, Pensions Panorama: Retirement-Income Systems in 53 Countries (The World Bank 2007) http://213.253.134.43/oecd/pdfs/browseit/8106101E.PDF

entitlement to a pension is achieved and how benefits are calculated. Pension systems variously strive to provide the elderly with a minimum level of income, to replace (at something less than 100%) the earnings a worker had prior to retirement, or to provide a standard payment to all senior citizens irrespective of other sources of income a pensioner might have. In most but not all highly developed nations, the public pension system combines more than one model; recipients of public pensions often but not always have access to private employment based pensions as well. In most nations, the public pension system is tied at least in part to an older person's work history and accounts for income available to the pensioner from other sources.

Means-tested social security systems are designed to ensure that all elderly persons receive a minimum monthly or annual income, irrespective of their past work history. An example of such a system is Australia's Old Age Pension. Under a second model, which is sometimes called the "basic" or "flat rate" pension, all persons of a certain "old" age (60 or 62 or 65) receive a fixed monthly payment, whether or not they have access to other sources of income (such as investments, savings, or a private pension). A variation of this model pays benefits on the basis of number of years worked (rather than on income earned through employment). Ireland, New Zealand, and Argentina have basic pension systems. A third model, represented by the United States' Social Old Age, Survivors, and Disability Insurance program, ties public pension distributions to a worker's inputs to the system. That is, retirees whose employment based income was higher during their years of employment receive higher payments than those whose incomes were lower. Replacement rates are lower for higher paid workers, but these workers' absolute pension payments may be two or three times more than those of lower-wage workers.

Most public pension systems are "pay-as-you-go", meaning that outlays – benefits paid to retirees-are supported primarily by the taxes or other inputs of current workers. As old-age dependency ratios increase, such systems become stressed as the taxes paid by workers begin to fall below the payments being made to beneficiaries. This has led to concerns in many countries about the long-term sustainability of the public pension system, and proposals for reform. But such proposals tend to focus only on assuring adequate funding for existing obligations, rather than on radical restructuring of pension systems to assure that the degree of income security for pensioners will not differ on the basis of gender.

How might a feminist approach to pension reform affect the nature and scope of that reform? The answer will depend both on the nature of the pension system, and on the nature of the feminist inquiry (liberal, cultural, radical) utilized. Although it is not possible in the short space of this chapter thoroughly to consider pension reform from a feminist perspective, the following examples illustrate the general approach.

Consider first the typical means-tested old age pension system, which assures a minimum income to persons, or married couples, who have attained a certain age. These systems generally do not distinguish between single men and single women in terms of the amount of the benefit. In some nations that have a means-tested pension, women become eligible for the pension at an earlier age than men.

Any reforms implemented as a remedy for declining pension reserves that assures that men and women are receive same minimum income and become eligible for that income at the same age (i.e., that equalizes the eligibility age for both women and men)[14] would be satisfactory under a liberal feminist approach, even if those reforms result in lower incomes for elders. Although potentially unjust from other theoretical approaches, reducing the incomes of both elderly women and elderly men does not offend the basic tenents of the liberal feminist approach to law and policy reform.

On the other hand, a radical feminist analysis of means-tested pension programs might look beneath the apparent "neutrality" of a same-pension-for-all persons system and ask "why are the great majority of impoverished elders women" (as is the case almost everywhere) and how can the pension system be utilized to correct the systemic problems that result in large numbers of older women in poverty? A radical feminist approach to pension reform might, therefore, preserve lower pension age-eligibility for women as an antidote to their more restricted access to employment opportunities over the course of their adult lives. Although such as system "privileges" older women vis-a-vis older men, in fact this "privilege" is required to correct the underlying discrimination against women in employment that has not yet been eradicated from the public sphere.

Public pension systems that are tied to a beneficiary's lifetime earnings from employment, such as the US Social Security system, are especially vulnerable to feminist critiques.[15] Under the US system, all workers pay a flat tax of 6.2% of earnings; employers contribute an additional 6.2% of earnings. This payroll tax is paid on the first $100,000 of income only. The tax is thus extremely regressive, and falls heavily on low-wage workers – a group disproportionately composed of women.

Even more problematic is the method of calculating benefits. A retiree's Social Security benefit is derived from the average of his or her earnings over the thirty highest-paid years of employment; the higher this average, the higher the retirement benefit. If a person has fewer than thirty years of documented, paid work, she is attributed with zero wages for the number of years needed to reach thirty; those zeros are included in the calculation of the mean. Thus, the system offers the greatest rewards those who have both worked in the public sphere all or most of their adult years, and who have the highest paid positions over the course of their employment history. And by including years in which a person received no wages in the calculation of lifetime average wage, this method of determining benefits severely penalizes persons who have not worked in the public sphere for at least 30 years.

It is virtually a truism that women earn less over the course of their lifetimes than men – because of pay discrimination in the same jobs that men have, because the kind of work women do is less valued, even when the work requires similar

[14] The UK, for example, will phase in an increase in the age the age at which women become eligible for the state pension from 60 to 65 (to bring in it in line with the age at which men are entitled) as a partial remedy for declining pension reserves.

[15] Information about the US Social Security system is available at http://www.ssa.gov/

education, training, and effort as the jobs men tend to do, and because as a group they are more likely to take time off from work than men to provide caregiving to dependent or ill relatives. Thus, any pension system that pays higher benefits in retirement to those workers who earned more wages during their working years will inevitably produce inequality between elderly men and elderly women. This inequality is, of course, exacerbated by the fact that men are more likely to have access to an employer sponsored pension than women. It is no wonder that the poverty rate in the United States among elderly women is almost twice that of elderly women.[16]

There are a number of ways that an inputs-based pension system can be made more gender-neutral. One option, which has been implemented on a relatively small scale in some Scandinavian countries, is to give credit towards retirement benefits for unpaid work that is deemed valuable to society – most notably, child-rearing. From the cultural feminist viewpoint, "accommodating" women's role as caregiver-nurturer is an appropriate way to acknowledge and value difference. A radical feminist approach to pension inequalities would entail more drastic measures, such as adjusting pension payments to account for women's lack of access to other sources of income and their socially-imposed obligation to withdraw from paid work for many years so that they can provide care, and by artificially adjusting the payments made to those who provide care to children and the elderly (as by imposing higher wage minimums for caregivers) through comparable worth theory. As a practical matter, of course, it is probably politically impossible to award sufficient "credit" for family caregiving, particularly for caregiving of the elderly, to account for the true value of such caregiving to societies.[17] The contribution of a feminist vision of pension reform is not necessarily to achieve full equality in one fell swoop, but instead to contribute to a broader discussion of the issues that acknowledge and account for the historic and continuing economic oppression of women at home, at work, and in their "golden years".

4.5 Conclusion

No single theoretical approach to elder law is capable of producing complete justice for all persons affected by law and aging policy. Aging affects individuals differently depending on many factors – economic status, gender, race, disability, and so on. It is important, therefore, that all perspectives, include feminist perspectives, be

---

[16] US Department of Health and Human Services, Administration on Aging, A Profile of Older Americans 2005, at 11, http://assets.aarp.org/rgcenter/general/profile_2005.pdf

[17] "The value of women's unpaid work is estimated to equal USD 11 trillion, or almost 50 percent of world GDP, yet this work is missing from national income accounts — leaving women missing out on social security, pension schemes and access to public services." United Nations Development Fund for Women, The Challenge: women are missing, and missing out, http://www.womenfightpoverty.org/challenge.php

included in policy discussions about law and aging and in legal texts that implement aging policy. Only then will "elder law" embody a rational and just account of the relationship between aging of the body and the positive law.

References

Bartlett KT, Kennedy R (eds) (1991) Feminist legal theory: readings in law and gender. Westview, Boulder

Campbell JC, Ikegami N (2000) Long-term care insurance comes to Japan, Health Affairs 26. http://content.healthaffairs.org/cgi/reprint/19/3/26.pdf

Cruikshank M (2003) Learning to be old: gender, culture, and aging. Rowman & Littlefield Publishers, Oxford

Dowd NE, Jacobs MS (eds) (2003) Feminist legal theory: An anti-essentialist reader. New York University Press, New York

Gilligan C (1982) In a different voice: psychological theory and women's development

Glickman H (2007) Financing long term care: lessons from abroad. Center for Retirement Research. http://www.bc.edu/centers/crr/issues/ib_7–8.pdf. Accessed on June 2007

Grimk S (1837) Letters on the equality of the sexes

Harris AP (1990) Race and essentialism in feminist legal theory. Stan. L. Rev. 42:581

Hiemstra G (2005) Futurist blog. http://futuristblog.blogspot.com/2005/01/ healthcare-and-elderly.html. Accessed January 14, 2005

MacKinnon C (1979) Sexual harassment of working women: A case of sex discrimination

MacKinnon CT (1983) Feminism, Marxism, method, and the state: Towards a feminist jurisprudence. Signs: J Women Culture Soc 8:635

MacKinnon C (1987) Difference and dominance: on sex discrimination. In: Feminism unmodified. Harvard University Press

MacKinnon C (1995) Speech, equality, and harm: The case against pornography. In: Lederer L, Delgado R (ed) The price we pay: The case against racist speech, hate propaganda, and pornography. Hill and Wang, New York

Pearson KC (2006) Finances, families and "Filial" laws: The real world as classroom, family focus. http://www.dsl.psu.edu/faculty/pearson/RealWorldEssayOnFilialLaws2006.pdf

Peterson PG (1999) Gray dawn: How the coming age wave will transform America – and the world. Crown

Screti F (2005) Elderly depend on immigrant women for caregiving. Global action on aging. http://www.globalaging.org/health/world/2005/depend.htm. Accessed on March 30, 2005

Williams WW (1982) The equality crisis: Some reflections on culture, courts, and feminism. Women's Rights L Rptr 7:175

Chapter 5
A Multi-Dimensional Model of Elder Law

I. Doron

5.1 Introduction

This book provides many theoretical approaches to understanding the field of law and ageing. However, unlike other chapters in this book, which try to provide a conceptual framework of the field through a single lens, this chapter proposes a different approach – a pluralistic one. This approach, does not try to argue that the uniqueness of the legal perspectives on ageing could be fully conceptualized through a single theoretical framework. By its nature, the pluralistic model which will presented in this chapter, argues that the only way to fully grasp the richness and diversity of elder law, is through a multi-dimensional model, which connects different functions and targets that law wishes to achieve.

The multi-dimensional model presented in this chapter is an on-going developmental process. The first version of the model was published in Doron (2003a). The model has been further developed, and an updated version was published in Doron (2003b). At the time, the model was presented under an Israeli-specific context. The goal of this chapter will be to further elaborate and develop this model, and provide it with a broad, international legal context. The multi-dimensional model can be graphically described as follows in Fig. 5.1 (Doron 2003b).

5.2 The Legal Principles Dimension

The center of the multi-dimensional model is embedded in general and universal legal principles. Every legal system has a core comprised of the system's underlying principles: the general, constitutional, and administrative norms that apply to any legal event in the given society. Such a core naturally includes the teachings of the constitution of the legal system as well as the basic principles and values underlying the various legal areas (Barak 1993; Raz 1972). As examples, one may consider the principle of equality before the law, drawn from the area of constitutional law; the principle of acting in good faith, regarding contract law; or the reasonability and proportionality principles, embedded in administrative law - all of which constitute

I. Doron (ed.) *Theories on Law and Ageing: The Jurisprudence of Elder Law*,

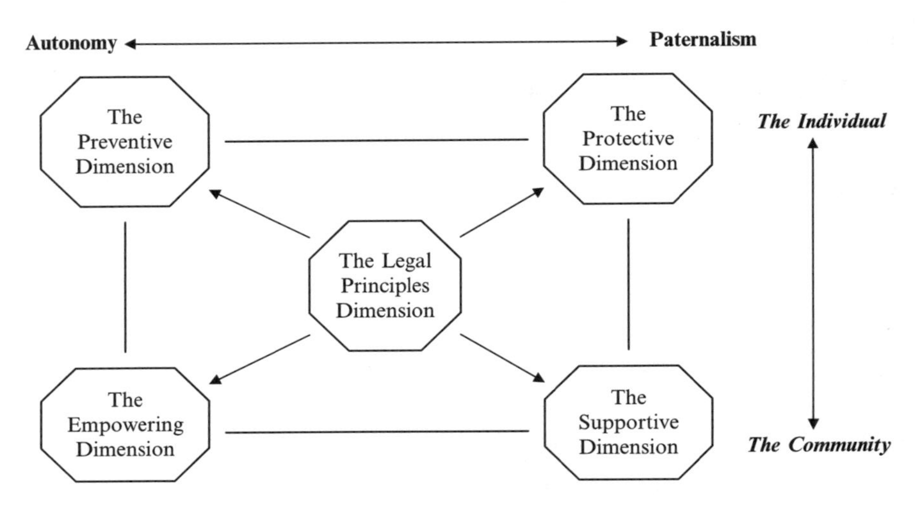

Fig. 5.1 The elder-law multi-dimensional model
(Source: Doron (2003b, p. 242–259))

part of the juristic core of many legal systems. The legal core protects the human and civil rights of the country's citizens, on a universal basis.

Historically, the existence of such a juristic core has supported the argument that there is no need to create a separate legal branch for dealing with the legal issues of the elderly. Legislators and the courts thought it unnecessary to create specialized norms for the elderly, on the assumption that these issues could be dealt with by means of the general legal principles that pertain to the entire population (Bonfield 1989). Thus, for example, a case of elder abuse and neglect was handled through general criminal law, which forbids administering physical harm to people. A case of financial exploitation of an elderly citizen was handled through the general directives and rulings pertinent to contract law, which forbid deception, exploitation, and coercion.

One legal example of this dimension is the legal principle of equality. Most modern constitutions include equality and the protection against discrimination as part of their central legal principles. For example, under the heading "Equality Rights," Sect. 15 of the *Canadian Charter of Rights and Freedoms* guarantees every individual the right to equal treatment by the state without discrimination. Age is one of the categories under Sect. 15 (Soden 2005, p. 264). The purpose of this legal principle is to prevent discrimination when people attribute stereotypical characteristics to individuals. This legal principle is also intended to ameliorate the position of groups, such as the older population, who have suffered disadvantage by exclusion from mainstream society.

Indeed, older persons have used this legal principle to defend and promote their rights. For example, in the province of Ontario, Canada, a legal challenge was taken against an Ontario health program which provided closed-circuit television magnifiers, but only to persons under 25 years of age. Per an appeal made by a 72-year-old man, who was refused this visual aid, the Ontario's Court of Appeal[1] viewed this age-based distinction as illegal. As the court noted, special programs should be designed so that the restrictions within the program are rationally connected to the objective of the program. The Appeal Court found that the program was initiated with age restrictions to have a small pool of clients and to conserve scarce financial resources, not because younger persons with disabilities have a greater need for such aids and less access to them than older persons (Soden at 296).

A different example, from Israel, also demonstrates the potential of the legal principles dimension to promote rights of older persons (this time, of older women). This is a well-known Israeli case known as the Dr. Naomi Nevo case (*Nevo v. The Jewish Agency* 1987). Dr. Nevo, a senior sociologist working for the Jewish Agency who was asked to retire at age 60, due to the Jewish Agency's collective agreement that mandated employee retirement for women at age 60 and for men at age 65. Dr. Nevo claimed that, as long as there was no one who repudiated her desire and ability to work until age 65, mandatory retirement at age 60 constituted blatant discrimination. The Supreme Court of Israel accepted her claim, stating that mandating retirement at age 60 only for women contradicted fundamental principles of the Israeli legal system. The Supreme Court based its verdict on a basic principle in Israeli law, the principle of equality. As the honorable Judge Bach described it:

> Discrimination is an affliction that creates feelings of deprivation and frustration. It has a detrimental affect on feelings of belonging and on the positive desire to participate in and contribute to society. A discriminating society is not a healthy one, and a country in which such practices are customary cannot be considered civilized.

Parallel with the court procedure, a legislative initiative took place in the Israeli parliament (Knesset), which concluded with the enactment of the *Employees' Equal Retirement Age Act* of 1987, according to which:

> Whenever a collective agreement determines that the retirement age for female employees is earlier than the male retirement age, a female employee will have the right to retire at any time between the retirement age set for female employees and that set for male employees.

The juristic core of the justice system has its advantages and its disadvantages when it comes to issues pertaining to the legal rights of the elderly. The advantage of the legal core in relation to these issues is twofold: its breadth transcends the boundaries of issues; its expanse is universal and pertains to all periods and to all areas of the law. In fact, this core provides legal protection to the elderly as it does to the entire population, regardless of the content of the case or its circumstances. The second advantageous aspect of this universal core is its implicit notion of normalization: the older population is perceived as an integral and equal part of society. The law treats the elderly as it does other parts of the population, and its rights are an integral part of civil rights as a whole. In and of itself, this egalitarian approach sends a positive message regarding the status of the elderly in society.

---

[1] Ontario (Human Rights Commission) v. Ontario (Ministry of Health) (sub nom. Roberts v. Ontario (Ministry of Health) (1994), 21, C.H.R.R. D/259 (Ont. C.A.).

However, this same claim also constitutes the disadvantage underlying the use of the juristic core as a legal tool for protecting the rights of the older population: the egalitarian principle cannot adequately contend with some of the particular issues that pertain concretely and specifically to the elderly. Unique social phenomena that have a distinct effect on the older population cannot be dealt with – or cannot be dealt with effectively - by means of this universal core. For example, findings from a recent Israeli commission on elder abuse recommended adding specific and new legislation in this field, since the existing penal law fails to address unique dimensions such as passive neglect, mental-neglect, or economic-exploitation (Commission on Prevention and Treatment of Elder Abuse and Neglect 2002). Thus, despite its advantages, this legal core is an insufficient tool, incapable of providing adequate legal responses to the needs of the elderly.

5.3. The Protective Dimension

The point of departure for this second dimension is indeed the need to address the unique and particular legal problems with which the older population must cope. The legal system responded by introducing point-specific legislation, intended for the most part to protect the elderly (among other especially weak groups) against negligence, poverty, and exploitation.

There are many legal examples for the protective dimensions of elder law. One example is the legal regime that assures quality of care in nursing facilities. As described by Frolik and Barnes (2003, p. 367), nursing home care presents special problems of quality assurance. Though the industry is increasingly professionalized, it is also subject to recurring incidents of notoriously poor care tied to high profit taking. Family members may have little power to assure quality and good value. Thus, both federal and state laws attempt to assure quality care and fairness in financial dealings with nursing homes, thus providing another legal layer of protection to institutionalized older persons. Such laws include diverse statutory measure such as licensing, periodical survey and certification, minimum staff, on-going training and more.

Another example for the protective legal dimension is the set of laws which attempt to combat elder abuse and neglect. In Israel for example, over the years various laws were enacted in an attempt to provide social protection for older persons. This resulted in different "legislative generations," as described by Doron et al. (2005):

5.3.1 The First Statutory Generation: Paternalism and Social Intervention

The first generation of laws relating to older people at risk of mistreatment was enacted, for the most part, during the 1950s and 1960s. During this period many laws were passed that granted relatively wide-ranging powers to the country's welfare authorities, including the authority to intervene in the lives of vulnerable persons, such as children, the retarded, and the insane (as they were named at the time). Two main laws enacted at that time are relevant for the aged: The Law of Legal Competence and Guardianship, 1962, and The Law for the Defense of Protected Persons, 1966- both of which enabled welfare officers to apply for guardianship or protective orders for older persons who could not care for themselves.

5.3.2 The Second Statutory Generation: Criminal Law and Mandatory Reporting

The second generation of protective legislation emerged at the end of the 1980s with the enactment of Amendment 26 to the Penal Code, 1989, had two principal aspects. The first is its explicit assertion that abusing "helpless persons," physically, mentally, or sexually, whether by omission or commission, is a criminal offence subject to severe punishment. The second major aspect of Amendment 26 was the establishment of mandatory reporting of any case, or suspected case, of abuse to an older and helpless person to a welfare officer or to the police. This obligation applied to all members of the public, but particular emphasis was placed on the fact that professional care workers were specifically mandated to report any abuse or neglect of older and helpless persons.

5.3.3 The Third Statutory Generation: Protection and Domestic Violence

In 1991 a completely new law was enacted: the Law for the Prevention of Violence in the Family, 1991. The aim of the new law was to provide temporary relief to victims of sexual, physical, and mental abuse within the family unit. Its chief innovation was the elimination of the approach that views welfare officers or the police as the solution to the problem of intra-family violence. The legislative innovation of the third generation was the adoption of a legal instrument that rested on civil law and could be set in operation quickly and independently by the victim or by a relative. The types of assistance provided by this legislation are mainly defensive and practical, and include such practices as physically removing the aggressor(s) from contact with the victim, prohibit the perpetrator to approach the victim in any manner, enter her or his apartment or even simply staying in a defined vicinity ("protective order") (Makies 1995).

The protective legal dimension provides the tools used daily by welfare officials, social workers, and a wide range of other professionals who provide care for the elderly. There is an abundance of examples in which these legal tools rescued elderly individuals from conditions of distress and abuse, and prevented damage to their health, property, and their independent functioning (Kerem 1995). The following case may serve as a typical example of how this protective legal perspective is applied in Israel. The health of a 77 year old woman who lived alone in her flat deteriorated such that she lost her sense of orientation and memory. She began wandering day and night without knowing where she was, injuring herself and risking her life and health in the process. Her only son lived abroad and was not in contact with her, and no one else in the community provided her with care. She was uncooperative with the local social welfare authorities and rejected attempts to provide her with in-home care. Under these circumstances, the welfare authorities appealed to the court to authorize them to move her to a proper institution against her own will, in accordance with the *Ward's Protection Act*. Moreover, the court was asked that a guardian be appointed to take care of her needs, according to the *Legal Capacity and Guardianship Act*. The court granted the request of the welfare authorities and made it possible to provide treatment and protection for this woman, by moving her to a nursing home, and thus preventing further deterioration of her health and welfare.

The advantage of the protective dimension is that, used wisely, it provides protection to those among the elderly who need protection. It is worth noting that findings, both in Israel and abroad, indicate that the problem of elder abuse is a serious social problem of an apparently broad scope (Lowenstein and Ron 2000; Ronen and Venikrug 1993; Decalmer and Glendenning 1997; Maclean 1995). With these findings in mind, the establishment of this legislative dimension was a necessary response to a real social problem.

However, there is criticism over the protective dimension. First, there are those who question the effectiveness of "protective" policies (e.g. criminal justice or "command and control" regulations) to actually make a difference in reality and provide real protection (Kapp 2000). Moreover, despite the phenomenon of elder abuse, it should also be noted that not all elderly individuals are weak, vulnerable, or subject to abuse. This state of affairs makes this legal dimension a double-edged sword: by treating the elderly paternalistically and making it possible to detract from their autonomy based on the ethical claim that the state has the right and obligation to intervene in one's life in order to protect the individual either from others or from him- or herself, the law substantiates the negative stereotype of the elderly population. This stereotypical approach stems from and reinforces a much broader social concept referred to as ageism, which in general means having a negative view of old age and the elderly (Bytheway 1995). In regard to this pervasive perception, this legislative dimension not only fails to protect, it actually causes additional detriment to the independence, freedom, and rights of the elderly (Doron 1999; Andrews 1997).

5.4 The Familial & Informal Supportive Dimension

The third legal dimension relating to the protection and promotion of elders' rights is directed in fact at a population other than the elderly themselves. This dimension reflects an awareness and understanding that, for a society that wishes to respect the rights of its elderly, attending to their legal status is not enough. Attention to the social context or setting is also warranted, particularly to the immediate circle of people who actually care for the needs of the elderly. This conviction was based on research findings that identified a correlation between quality of life, health, and the ability to age in a respectful manner and the involvement of informal, social support networks. The elderly who had a strong supportive network, such as that of nearby family members who felt obligated to provide for and support them, were less likely to be ill, placed in institutions, or to have a guardian appointed to ensure their proper care (Kane and Penrod 1995).

To continue to help and assist social support networks for the elderly, a variety of legal tools have been developed both in Israel as well as worldwide. First, laws were passed to enable relatives who act as ongoing informal caregivers of elderly family members to minimize the conflict between their obligations to the workplace and their burden of care. Second, statutory frameworks were created to provide monetary support for relatives who act as informal caregivers of elderly individuals, to help compensate for expenses and financial losses incurred as a result of this role. Finally, formal networks were established to provide temporary or continuous nursing or palliative care, in an attempt to complement or facilitate (but not as a substitute for) the informal care (Schmid and Borowski 2000; Gerald 1993).

Taking England as an example, since the 1990, there is evidence for legal awareness of the importance of 'family friendly' employment policies in the work place and their vitality to family caregivers. Section 8 of the Employment Relations Act 1999 has given employees the right to a reasonable amount of time off as unpaid leave to provide assistance to "dependents" or to deal with unexpected disruption to their care (McDonald and Taylor 2006, p. 36). A different interesting legislative development in England touched upon the legal recognition of the status of "carer." Thus, the Community Care Act 1990 included a wide definition of carers as "families, friends and neighbors" who support vulnerable people, and whose preferences should be taken into account in any community care assessment of the person for whom they care (McDonald and Taylor 2006, p. 34). Further legislative developments, such as the Carers (Equal Opportunities) Act 2004, require social services assessor to inform carers of their right to request an assessment and also requires the assessor to take into account the carer's wish to undertake employment or pursue education (p. 35).

The advantage of legally protecting elder rights by supporting those who care for them is derived from the link between the legal tool and the findings of other, extra-legal, scientific disciplines, such as gerontology, sociology, and psychology. Thus, the non-legal disciplines are instructive, as they demonstrate the significant role played by the informal social support networks in ensuring the rights of the older population and creating a more positive image of the elderly. Although also this dimension considers the elderly a weak group in need of support and assistance, it calls for providing informal social support networks, rather than external, unwanted, paternalistic state interventions.

Nevertheless, here too, the moral and social strength of this legal dimension is not without limitations and apprehensions for those involved. First, there are doubts regarding the ability of such legal tools to influence the existence of informal support networks. Some findings show that providing formal support does not change the nature of the informal relationships (Horowitz and Shindelman 1983; Arling and McAuley 1983). Second, there are also political and moral reservations as to the justification of legal involvement in this regard: Does this dimension enable the State to relinquish its obligation to take care of its elderly? Or, conversely, aren't the issues addressed in this dimension part of a personal moral obligation to the elderly that should not involve the law? Finally, there are those who cast doubt on the financial astuteness of this legislative dimension, claiming that in a limited-resource environment, it may be more appropriate to finance the formal rather than the informal support systems for the elderly (Naon et al. 1992; Nelson and Nelson 1992).

5.5 The Preventive Dimension

The fourth legal dimension that relates to the status of the elderly population concerns legal ways to prevent unnecessary intervention in the lives of the elderly by means of creating legal planning tools.[2] At this dimension, elder law aims not only to protect the elderly or those who care for them, but directs its focus to the abilities of the elderly themselves. This focus is expressed in the formulation of legal tools that enable the elderly to design their own lives, control decisions pertaining to a time when their physical and mental conditions may deteriorate, and thus preclude outside legal intervention. These tools are relevant both to property and health care, and are relevant both to old age, and even after death.

In the field of income security in old age, private pension schemes (e.g. the 401(k) plans) are an important planning tool. Any decision regarding participation and contribution to such plans has critical effect on the retirement income of older persons (Paterba 2005). Another, more recent economic planning tool is the reverse-mortgage (or other home-equity conversion instruments such as sale-leaseback), which allow "house rich cash poor" elderly to receive additional monthly income (Hammond 1993). In the field of property-management and estate-planning there is a wealth of legal institutions that provide a diversity of economic and legal options that will ensure that one's wishes concerning his or her own property, will be honored in case on incompetence or loss of personal capacity to manage one's life. These instruments include legal tools such as durable powers of attorneys or various trust accounts.

A significant financial issue that also needs planning and prevention in old age is the cost of institutional long-term care (or home-based long term care). In recent years, there is growing awareness to personal long-term care insurance policies.

---

[2] Various aspects of this legal dimension are discussed in length in this book by Prof. Lawrence Frolik. See Chap. 2.

Such individual insurance policies may be the financial answer, at least to some older persons, to prevent catastrophic out-of-pocket costs (Wieneret al. 2000). For those who try to evade institution by informal care-giving, a whole new set of contractual planning has been developed in recent year: the private-care agreement. A private care agreement usually involves a transfer of property (usually the family home) to a friend or family member in exchange for a promise of care and support – an arrangement that create a whole range of potential legal issues (Hall 2002). Finally, a whole set of legal planning issues surround older persons entering in what is known "later-in-life marriages," starting from tailored prenuptial agreements, agreements to avoiding medical obligations and protecting pension income to inheritance and children's rights (Grama 1999).

In the field of health care decision making and issues that surround "death with dignity," law has created in the last two decades various new legal instruments. From living will, to advance directives, to continuing powers of attorney for health care – all these novel legal documents allow a person to decide in advance, which (and if) medical treatment will be administered to him or her once incapable or in the end stages of life.

When planning to "after-death," once again, law provides different legal planning tools. A familiar example of such a legal planning tool is the will. This is an ancient legal tool that enables individuals to plan what will become of their property following their death and to express in this manner their independence and control of their estate. Another example is the ability to decide in advance to donate your organs (for transplantation) or even your body (for scientific research). In many countries, law enable a person to also plan one funeral plans, where and how one wishes to be buried, and enter what is known as "preneed funeral contracts" (Frank 1996).

Certainly, the greatest advantages of this dimension are its individualistic and autonomous perspective as well as its preventive dimension. Its point of departure is that the elderly themselves should have the ability to design the last decades of their lives and define how their rights should be protected. This acknowledges the significance of their independence, self-respect, and the value of maintaining their personal autonomy. Moreover, these legal planning tools make it possible to avoid the need – on a social scale – to resort to the formal judicial system and its drawn-out adversarial procedures. The personal planning tools are essentially flexible, in that they can be easily adjusted to correspond to the particular needs and desires of the individual. They are also potentially more sensitive to ethnic, cultural, and moral differences, in that they do not prescribe a uniform result, but rather render decisions that are based on individual values.

However, it seems that the main weakness of this preventive dimension is the gap between its theoretical rigor and its actual implementation as experienced by many older persons. The application and use of these planning tools requires a combination of awareness of and familiarity with this set of tools, as well as financial and intellectual know-how. Many elders are unaware of these tools, many may not have the financial means to acquire them and, in addition, there are psychological and social difficulties associated with the sensitive topics involved. Thus, in fact, these tools remain, in various degrees, unused. In the past, several American studies had confirmed these apprehensions, indicating that the rate of use of these tools among the elderly population was indeed low (Larson and Eaton 1997) and when they were used, all too often the medical team disregards these legal documents (Teno et al. 1994). While this reality have been improved in the US (AARP 2000), in other countries, e.g. Israel, the rate of usage is still low, and there is still a significant gap between the general awareness and actual usage of different legal planning tools for old age (Doron and Gal 2006). Even if it were possible to overcome the issues of awareness and faulty implementation, the moral and social basis of this dimension may be questionable. This preventive dimension of personal planning tools is founded on a philosophy of the individual's rights, pertaining specifically to the social strata of the elderly, many of whom focus their lives around dimensions of support, assistance, and dependence. It has been suggested that for many elders, individualism and independence are less significant factors in their daily lives and less essential to their daily functioning than reciprocity and assistance. This claim should not be treated lightly: environmental and community involvement obviously is not compatible with legally supported actions based solely on the individual's wishes. Thus, this tool may entail the risk of isolating older persons, if it ignores the desires, interests, and worries of family members, professionals, and others involved in caring for the elderly individual (King 1996; Fraser and Gordon 1997).

5.6 The Empowerment Dimension

The final legal dimension relating to the older population was formed once legal systems came to realize that guarding the physical and financial welfare of the elderly was insufficient, as were the individual planning tools, which remained, for the most part, unused. Research in all the previously mentioned dimensions found unawareness, lack of knowledge or low usage and takings of existing rights (Arksey et al. 2000; Doron and Werner, in press). Without legal means that could initiate a movement toward change, progress, and realization, the previous dimensions of legal tools would not accomplish their intended goals.

This understanding was based on an observation of the current reality, which revealed that adding new legislation to the law books did not influence the real life, daily experiences of the elderly. Older persons would not be able to realize their legal rights as long as they remained unaware of them, uninformed about their purpose, and unassisted in implementing them. Various laws, which intended to improve the status of the elderly, could not be put to use without appointing someone to oversee their proper enforcement. This perspective led to the recognition that legal support for the empowerment of the elderly population was the step necessary to initiate a change in the elderly population's social and political status, and that this would help attain the goals of all the prior legislative dimensions (Thursz 1995).

Empowerment as a theoretical term is complex and implies various definitions and approaches, the scope of which exceeds the limits of this article. However, it is worth noting that for our purposes, empowerment of the elderly attempts to address the political asymmetry in power distribution and particularly the weak social status of the elderly, by both increasing their power and changing the social environment in which they function (Handler 1995).

One example of an empowering legal policy is the Older Americans Act subsidized legal assistance program (Frolik & Barnes 2003, p. 79). Also known as "Title IIIB" funding, the OAA identified legal services as a priority area that must receive an "adequate portion" of funding by each Area Agency on Aging administering OAA funds. While these legal services target older individuals with economic or social needs, unlike other legal aid services, they are not subject to means-testing, and cover not only the poor. The rationale of these services is an empowering one: providing legal information, advocacy and representation, will allow older people to be more aware of their rights, enable them to exercise existing legal rights, and eventually enforce greater respect for older people's rights.[3]

Another example of an empowering legal policy is mandatory representation of older person in guardianship proceedings. In light of the far-reaching legal consequences of nominating a guardian over an older person, many states in the US require that the respondent be represented by either legal counsel or a guardian-ad-litem at the hearing. Moreover, if the respondent cannot afford counsel the state will bear the cost (Frolik and Barnes 2003, p. 469). By providing legal counsel, one can overcome the power imbalance, and ensure that older persons subject to guardianship proceeding are indeed granted due process of law.

The legal dimension of empowerment is well accepted by those interested in promoting elders' rights. However, it is also the focus of criticism that is far from insignificant. The first claim against this dimension is that its point of departure defines the elderly once again as a powerless group, in contrast to the powerful group that is either willing or forced to "grant" or "transfer" some of its powers. This perception adopts the negative or stereotypical attitude toward the elderly (Sykes 1995). The second critical claim casts doubt on the efficacy of empowerment tools as a means for changing the social status of the elderly or of any weak social group, in general (Weissberg 1999). Thirdly, there are concerns about the enforceability and actual ability of empowering legislation to make a difference, especially in light of the political weakness of the elderly population (Brick 2002). Indeed, at least one study proved that the implementation and usage of the Israeli *Senior Citizens Act* is limited (Senior Citizens Discounts 2002). Finally, it is claimed that the empowerment of the elderly usually comes at the expense of another social group that becomes disempowered. This dynamic may render a result that is contrary to the original intention: instead of empowering the elderly it

[3] It should be noted that there is some criticism over the OAA legal services, e.g. regarding the limitations placed on reform activities such as the prohibition of participation in a class action suit or a political lobby/advocacy for particular public policies (Frolik and Barnes 2003, p. 81).

places them in conflict with other social groups vying for power in society. Often these competing groups are in fact stronger than the elderly, and thus, instead of improving the social status of the elderly, this approach results in further detriment to their legal status (Lightman and Aviram 2000).

5.7 The Broader Picture

5.7.1 The Diversity and Variability of the Field of Elder Law

The multi-dimensional model introduced here affords a clear presentation of the diversity and variability that characterizes the field of elder law. Elder law, so the multi-dimensional model suggests, includes distinct types of legal tools, a range of political and philosophic approaches, and multiple perspectives on the concept of "elder rights." In other words, this field is by no means directed by a single viewpoint: it cannot be considered either "individualistic" or "paternalistic," nor can it be portrayed as promoting "negative" as opposed to "positive" rights, or as favoring the individual over the family, the private over the public. Elder law covers the range of possible approaches and perspectives, and it can be fully appreciated only by understanding the internal balance between all of its components.

The model also encompasses two central dichotomies that serve as a key to many discussion within elder law. The "paternalism" vs. "autonomy" dichotomy, is represented on the one hand by the protective and family-support dimensions, which both represent societal willingness to provide care for the elderly, while sometime paying the price of limiting the autonomy or freedom of choice. On the other hand, the preventive and empowering dimensions stress the value of personal autonomy, and willingness to give legal power to one's personal preferences, even if such preferences are against his or her "best interest," family-members preferences, or societal moral. Another important dichotomy is the "individual vs. societal" dichotomy: while both the protective and preventive dimension work mostly on the individual levels, the family-support and empowerment dimensions work mostly, or more strongly, on the broader community.

5.7.2 The Significance of the Whole

The multi-dimensional model makes it possible to understand that a specific law or a concrete legal position regarding elder rights is in fact only one component of a much larger framework whose various parts complement each other. A discussion of the personal planning tools dimension that ignores the dimension of empowerment tools may lead to inaccurate conclusions. Likewise, familiarity with the protective legal dimension laws without taking into account the dimension of supportive network tools renders an incomplete understanding of the issue of elder rights as a whole. Only an examination of the various legal dimensions in the context of the entire picture produces an accurate and helpful assessment of each legal dimension individually.

5.7.3 *The Openness of the Model*

The model allows for more dimensions to be added, with the understanding that this is not a closed list, but rather an open framework composed of a variety of fundamental approaches to the issue of maintaining elder rights. It is only natural that scholars should seek to add new dimensions, and it is conceivable that the overall goal of securing elder rights will be achieved in the long-term. Also conceivable is a description of the existing reality based on legal dimensions other than those presented here. In this sense, the multi-dimensional model is still in the process of formation, and as such, should be considered an invitation to extend the discussion on these issues.

5.7.4 *The Significance of "Choice"*

The model demonstrates that older persons and those who care for them have the option, in most cases, to choose from among a variety of legal paths. In other words, today's legal reality is such that there are rarely cases in which the "default" mode is a single legal approach. On the contrary: in the majority of cases, several legal tactics are possible, from amongst which the most suitable approach can be selected. For example, in an attempt to respond to a potential elder abuse situation, one may choose to utilize the penal code and file a complaint to the police; one might prefer to educate and empower the potential elderly victim by providing information about the various legal recourses available; or alternatively, one may attempt to neutralize the social source of abuse by providing legal and financial support to the informal care-taker, who might be the potential abuser. Thus, the final choice is always a function of values and preferences rather than a prescribed or statutory necessity (Kapp 1996).

5.7.5 *The Model as a Tool for International Comparative Legal Study*

The application and implementation of the multi-dimensional model presented here can be pertinent to different legal systems and be used to examine any legal system or to analyze its various laws by observing how these correspond to each of the model's suggested dimensions. For example, one can focus on a single dimension and compare various components of its legal content in different legal systems (as was done in Lechner and Neal (1999), regarding the supportive dimension), or one may even narrow the comparison further, to a single element within one dimension (as was done in Doron (2002), per elder guardianship). In this manner, either a particular dimension or the multi-dimensional model as a whole can be a used as an efficient comparison tool in international and comparative law.

5.8 Conclusion

The plurality of theoretical perspective described in the various chapters in this book demonstrates the challenge facing any one who attempts to provide an holistic approach to elder law. Within this broad context, the proposed multi-dimensional elder-law model offers a suitable framework for establishing a rich social and cultural interaction between law and ageing. The model embodies the view that a genuine commitment to empowering older persons and ensuring their rights and freedoms is possible only through a holistic approach: one that cares for the elderly at home, in institutions, at the workplace, and in retirement; one that allows protective services when vulnerable or weak, and provides information, knowledge, advocacy, and end-of-life care with dignity. The multi-dimensional model mandates an approach to human rights issues of the older population, which goes beyond any specific or single-lens legal analysis. It provides a broad framework that respects cultural differences as well as diversity in legal traditions, and promotes cooperation between government agencies, communal and non-governmental organizations, and family members. Finally, from a theoretical perspective, it mandates a pluralistic vision, which enables us to escape the illusion that we can fully understand the field of law and ageing from any singular approach.

References

AARP. (2000). *Where There Is a Will… Legal Documents Among the 50+ Population: Findings From an AARP Survey*. Washington D.C.: AARP.

Andrews MD (1997) The elderly in guardianship: a crisis of constitutional proportions. Elder Law J 5:76–105

Arksey H, Hepworth D, Quereshi H (2000) Carer's needs and the carers act. Social Policy Research Unit, University of York, New York

Arling G, McAuley WJ (1983) The feasibility of public payments for family caregiving. Gerontologist 23(3):300–306

Barak A (1993). Interpretation of the law volume 2: interpretation of legislation. Nevo, Tel-Aviv [in Hebrew]

Bonfield L (1989). Was there a 'third age' in the pre-industrial English past? Some evidence from the law. In Eekelaar J, Pearl D (eds) An aging world – dilemmas and challenges for law and social policy. Clarendon Press, Oxford, pp 37–53

Brick I (2002). The politics of aging. Eshel, Jerusalem [in Hebrew]

Bytheway B (1995) Ageism. Open University Press, Buckingham

Commission Report on Prevention and Treatment of Elder Abuse and Neglect (2002) Eshel, Jerusalem [in Hebrew]
Decalmer P, Glendenning F (1997) The mistreatment of elderly people. Sage, London
Doron I (1999) From lunacy to incapacity and beyond: guardianship of the elderly and the ontario experience in defining 'legal incompetence'. Health Law Can 19(4):95–123
Doron I (2002) Elder guardianship kaleidoscope: a comparative legal perspective. Int J Law Policy Family 16(3):368–398
Doron I (2003a) Law and geriatrics: an Israeli perspective on future challenges. Med Law 22(2):285–300
Doron I (2003b) A multi-dimensional model of elder law: an Israeli example. Ageing Int 28(3):242–259
Doron I, Alon S, & Offir N. (2004). Time for policy: Legislative response to elder abuse and neglect in Israel. *Journal of Elder Abuse and Neglect*, 16(4), 63–82.
Doron I, Gal I (2006) The emergence of preventive elder law: an Israeli example. J Cross-Cult Gerontol 21:41–53
Doron, I. & Werner, P. (in press). Facts on Law and Ageing Quiz. *Ageing and Society.*
Equal Retirement Age for Female-Worker and Male-Worker Act of 1987
Frank JA (1996) Preneed funeral plans: the case for uniformity. Elder Law J 4(1):2–54
Fraser N, Gordon L (1997) A genealogy of "dependency" – tracing a keyword of the U.S. welfare state. In Fraser N (ed) Justice interruptus – critical reflections on the "postsocialist" condition. Routledge, New York, pp 121–150
Frolik LA, Barnes AP (2003) Elder law: cases and materials. LexisNexis, Newark
Gerald LB (1993) Paid family caregiving: a review of progress and policies. J Aging Soc Policy 5:73–89
Grama JL (1999) The "new" newlyweds: marriage among the elderly, suggestions to the elder law practitioner. Elder Law J 7(2):379–407
Hall M (2002). Private care agreements between older adults and friends or family members. CCELS, Vancouver
Hammond, C. M. (1993). Reverse mortgages: A financial planning device for the elderly. *The Elder Law Journal*, 1, 75.
Handler JF (1995) Community care for the frail elderly: a theory of empowerment. Ohio State Law J 50:541–593
Horowitz A, Shindelman LW (1983) Social and economic incentives for family caregivers. Health Care Financ Rev 5(2):25–33
Kane RA, Penrod JD (eds) (1995) Family caregiving in an aging society – policy perspectives. Sage, California
Kapp MB (1996) Enhancing autonomy and choice in selecting and directing long-term care services. Elder Law J 4(1):55–81
Kapp MB (2000) Quality of care and quality of life in nursing facilities: what's regulation got to do with it? McGeorge Law Rev 31:707–731
Kerem B (1995) Protecting the elderly. Ministry of Labor & Welfare, Jerusalem [in Hebrew]
King NP (1996) Making sense of advance directives. Georgetown University Press, Washington
Larson EJ, Eaton TA (1997) The limits of advance directives: a history and assessment of the patient self-determination act. Wake Forest Law Rev 32:249–293
Lechner VM, Neal MB (eds) (1999) Work and caring for the elderly. Brunner/Mazel, Philadelphia
Lowenstein A, Ron P (2000) Elderly abuse by caring relatives: frequency of the phenomenon in Israel, typology of the victim and the abusing relative, and etiology of the abuse. Soc Welfare 20(2):175–198 [in Hebrew]
Lightman ES, Aviram U (2000) Too much, too late: the advocacy act in ontario. J Law Soc Policy 22(1):25–48
Maclean MJ (ed) (1995) Abuse & neglect of older canadians: strategies for change. Thompson, Ottawa
Makies R. (1995). The Law of Prevention of Domestic Violence. In F. Radai, C. Shalev & M. Leiben-Kobi (Eds.), *The Status of Women in Law and Society* (pp. 307-317). Tel-Aviv: Shoken [in Hebrew].

McDonald A, Taylor M (2006) Older people and the law. The Policy Press, Bristol
Naon D et al (1992) Preventing the institutionalization of disabled elderly by providing ongoing nursing care through the community. Joint, Jerusalem [in Hebrew]
Naomi Nevo V. (1987) The Jewish Agency & the National Labor Court, Supreme Court Rulings (Piskei Din) 44(4), 749
Nelson HL, Nelson JL (1992) Frail parents, robust duties. Utah Law Rev 3:747–763
Paterba JM (2005) Individual decision making and risk in defined contribution plans. Elder Law J 13(1):285–308
Raz J (1972) Legal principles and the limits of law. Yale L J 81:823–854
Ronen M, Venikrug S (1993) Understanding the issue of elder abuse in Israel. Soc Welfare 14(4):17–32 [in Hebrew]
Senior Citizens' Discounts (2002) Baduk – Israel's consumer council magazine, 43, 1–3 [in Hebrew]
Schmid H, Borowski A (2000) Selected topics in the matter of community-based long-term care services to the elderly a decade after implementation of the law. Soc Secur 57:59–71 [in Hebrew]
Sodon A (2005) Advising the Older Client. Toronto: Butterworths.
Sykes JT (1995) A second opinion. In Thursz D, Nusberg C, Prather J (eds) Empowering older people: an international approach. Cassell, London, pp 47–59
Teno JM et al (1994) Do formal advance directives affect resuscitation decisions and the use of resources for seriously ill patients? J Clin Ethics 5:23–30
Thursz D (1995) Introduction. In Thursz D, Nusberg C, Prather J (eds) Empowering older people: an international approach. Cassell, London
Wiener JM, Tilly J, Goldenson SM (2000) Federal and state initiatives to jump start the market for private long-term care insurance. Elder Law J 8(1):57–102
Weissberg R (1999) The politics of empowerment. Praeger, New York

Chapter 6
A Law and Economics Approach

R.L. Kaplan

6.1 Overview of Law and Economics

No less an authority than Yale Law School Professor Bruce Ackerman has dubbed the economic approach to law "the most important development in legal scholarship in the twentieth century" (Cooter and Ulen 2000, p. 2). But what exactly is this approach? The leading textbook in the area contends that "[e]conomics provide[s] a scientific theory to predict the effects of legal sanctions on behavior" (Cooter and Ulen 2000, p. 3) and that "economics provides a behavioral theory to predict how people respond to changes in laws" (Cooter and Ulen 2000, p. 3). Economics is able to accomplish this important function, its proponents assert, because it "has mathematically precise theories (price theory and game theory) and empirically sound methods (statistics and econometrics) of analyzing the effects of prices on behavior" (Cooter and Ulen 2000, p. 3).

At the same time, these authors claim that "economics provides a useful normative standard for evaluating law and policy" (Cooter and Ulen 2000, p. 3) – namely, efficiency. This value is clearly important because, as the authors note rather convincingly, "it is always better to achieve any given policy at lower cost than at higher cost" (Cooter and Ulen 2000, p. 4). Notions of fairness, or "distribution," however, are less amenable to economically informed analysis, as the authors admit (Cooter and Ulen 2000, p. 4). Nonetheless, the premise of law and economics is that such analysis is an important, if not a superior, lens through which to evaluate legal rules in terms of the policies that they purport to promote.

One of the undisputed gurus in this area, Judge Richard A. Posner, formerly a professor at the University of Chicago Law School, is even more unabashed in his advocacy of the law and economics approach. In his book entitled *Aging and Old Age* (Posner 1995), he applies the law and economics approach to a variety of issues surrounding aging generally and elder law in particular. He does so because he believes that "economics can do a better job of explaining the behavior and attitudes associated with aging, and of solving the policy problems that aging presents, than biology, psychology, sociology, philosophy, or *any other single field* of natural and social science" (Posner 1995, p. 2 (emphasis added)). That level of self-confidence bordering on hubris characterizes much of the law and economics literature, notwithstanding the tentativeness of many of its analytical mechanisms.

I. Doron (ed.) *Theories on Law and Ageing: The Jurisprudence of Elder Law*,

6.1.1 Economics of Physician-Assisted Suicide

Judge Posner moves beyond such seemingly natural subjects for his approach as retirement savings incentives and insurance regulation into physician-assisted suicide for the elderly. In this context, he contends that the dying process is generally characterized by serious pain, fear, and confusion (Posner 1995, p. 239) and that terminally ill patients can rationally conclude that the benefits of choosing a dignified death exceed the costs of doing so. He refutes the claim that this approach devalues life by observing that "[t]he spectacle of nursing homes crowded with frail and demented old people, or of hospital wards crowded with dying people so heavily sedated as to be barely sentient or so twisted with pain as to be barely recognizable, might be thought rather to undermine than to enhance a sense of the preciousness of life" (Posner 1995, p. 241). Moreover, Posner explains, society ought to empower physicians to assist prospective candidates for suicide, because the availability of such professional assistance is likely to "lower the cost not only of suicide but also of interventions that can avoid suicide" (Posner 1995, p. 249). What distinguishes this discussion, it should be noted, is not so much the conclusion as the mode of analysis, complete with algebraic formulae (see Posner 1995, pp. 246–247).

6.1.2 Economics of Friendship Formation Among the Elderly

Judge Posner extends his doctrine of economics *über alles* to suggest that older people are less likely to form new friendships because the economic cost of investing in a new relationship is too large in terms of the anticipated benefits that such a relationship would yield. This "human capital" approach certainly has some intuitive appeal, but it necessarily assumes that everyone – and especially all older people – utilizes strict cost-benefit calculations in determining their every move. Even as sympathetic a fellow traveler in the law and economics arena as my colleague Thomas Ulen felt compelled to express some skepticism about Posner's analysis:

> "… I much doubt that decisions to form acquaintanceships at an advanced age are really strongly influenced by considerations like those of human capital theory. There are other factors that strike me as much more important in that decision, such factors as where one happens to live (whether in a retirement community or alone, in a town where one has long lived or in a new locality, near one's children and grandchildren), whether one is married or not, whether one is active in clubs or other informal groups, and of course, one's personal tastes." (Ulen 1996, p. 112 (reviewing Posner RA 1995, Aging and old age))

As the younger generation might say more succinctly, "well, duh." Yet, it is precisely this sort of injection of common sense and basic intuition that is needed to counter the cold determinism that often results from the blind adaptation of economic theory to human decision-making and other interactions.

6.1.3 *Sampling of Economic Elder Law Issues*

That is not to say, however, that the law and economics approach has little jurisprudential value. Very much to the contrary, economically informed analysis is critical to understanding an enormous range of issues that are at the core of elder law, both on a policy level and at the level of individual elders. A quick sampling of these issues might include the following:

- Does the existence of Social Security lower some people's propensity to save for their retirement?
- Does this program encourage inappropriately early withdrawal from the compensated workforce?
- Does the availability of Medicaid coverage for nursing home expenses diminish people's willingness to obtain private long-term care insurance? Does such coverage encourage families to place older relatives in nursing homes earlier than they would if this coverage did not exist?
- Is coverage of long-term care costs more efficiently borne by a broad-based system of social insurance than by individually secured policies with their higher distribution expenses?
- If families were responsible for older relatives' health care expenses, would elder abuse cases increase in frequency?
- Why do most retirees avoid annuities even though these contracts ensure that their purchasers' accumulated retirement savings will last as long they do? Can annuities be made more economically appealing despite the significant adverse selection problem with respect to likely purchasers?
- Does the Medicare program's guaranteed enrollment discourage pre-retirees from following healthy lifestyle choices regarding smoking, food selection, and exercise?
- Can a voluntary prescription drug insurance program effectively counter the problem of adverse selection; i.e., will only elders who anticipate being heavy consumers of prescription medications enroll in the program?
- If the government were to pay for long-term care in patients' homes, would such coverage create moral hazard problems with family members who would otherwise provide such care themselves?

6.2 Application Of Law and Economics to Two Major Elder Law Issues

According to Judge Posner's definitive treatise entitled *Economic Analysis of Law*, "The 'economic theory of law'… tries to explain as many legal phenomena as possible through the use of economics" (Posner 2003, p. 26). To illustrate some of the insights and caveats that this approach can produce, this chapter takes an extended look at two of the more frequently encountered issues in elder law:

(1) Should older homeowners tap the equity in their residences via reverse mortgages?
(2) When should a person begin taking Social Security retirement benefits?

These two sections necessarily reference the US legal system regarding elder law, but similar considerations apply elsewhere to greater or lesser degree. Those differences, to be sure, might be extremely significant. For example, the financing of health care in retirement presents very different issues in a country with a national health care delivery system or some variant of universal access to health care. Similarly, countervailing societal considerations such as China's encouragement of retirement at age 60 for men and age 55 for women (see Heller 2006, p. 20) pose palpably different dilemmas than are present in the United States with its panoply of alternative arrangements.

6.2.1 Reverse Mortgages for Older Homeowners

Fully four out of five households headed by an older person in the United States own their primary residence and in nearly three out of four such circumstances, there is no mortgage debt outstanding on their property.[1] For the majority of older households, moreover, the primary residence represents their largest single financial asset and in many parts of the country, an asset that has seen significant price appreciation during their period of ownership. But many such homeowners need regular infusions of cash to supplement other sources of retirement income, to pay for recently increased medical expenses, to bestow gifts on favored family members, and for various other purposes. The dilemma that these elders face is how to tap the equity they have accumulated in their primary residence without selling the property and moving out. Such a move is traumatic at any age but particularly for older people who have lived in their home for several decades already and want to continue doing so. Indeed, a survey of older homeowners by AARP (formerly the American Association of Retired Persons) found that fully 86% wanted to remain in their home until they passed away (Frolik and Kaplan 2006, p. 189).

To accommodate these people, a financial instrument known as a "reverse mortgage" has been developed whereby a bank or other financial institution pays a fixed sum every month to the homeowner as long as that homeowner, or homeowners if the property is jointly owned, lives in the residence (see generally, Frolik and Kaplan 2006, pp. 209–216). These arrangements are denominated "reverse mortgages" to distinguish them from the more familiar variety whereby homeowners make monthly payments *to* a financial institution. Reverse mortgage payments accumulate, along with accrued interest, and must be repaid when the homeowner(s)

[1] Administration on Aging, U.S. Dep't of Health and Human Services, A Profile of Older Persons: 2005 11 (2006), available at http://assets.aarp.org/rgcenter/general/profile_2005.pdf (last accessed 5 July 2007).

leaves the property – whether that departure is occasioned by the person's death or a person's move to a medical facility or some relative's home. But until that time, the bank pays a monthly sum determined by the residence's market value, prevailing interest rates, and the homeowner(s)'s age. These monthly payments, moreover, can be used for any purpose whatsoever, including in-home nursing assistance, premiums on long-term care nursing home insurance, prescription medications, gifts to relatives, utility bills, annual property taxes, and the like (see generally, Frolik and Brown 2000, paragraph 16.08; Regan et al. 2007, paragraph 12.05; Scholen 1993; Reilly 1997; Hammond 1993).

6.2.1.1 Applicable Legal Regimes

Various federal and state laws limit the availability of reverse mortgages to homeowners who are of a certain age (most commonly 62 years old)[2] and often mandate financial counseling by some entity other than the financial institution that would be originating the reverse mortgage.[3] The question that this section now considers is how the tax law affects homeowners who might be considering a reverse mortgage. This question, in turn, has three components: (1) taxability of the loan proceeds, (2) deductibility of the interest, and (3) taxation of gain on ultimate disposition of the residence.

The proceeds of a reverse mortgage are treated the same as proceeds from any other kind of loan – namely, no tax consequences because the obligation to repay these proceeds eliminates any notion that the amount received represents an increase in the borrower's net worth (see Bittker et al 2002, paragraph 4.01, p. 4–3). The interest expense that accrues on a reverse mortgage provides no tax deduction, even though it accrues in reference to the homeowner's principal residence.[4] This expense is not paid, after all, until the reverse mortgage becomes due, so there is no tax deduction until that time. Even then, only the interest on the first $100,000 of a reverse mortgage's principal amount owed gives rise to a tax deduction,[5] but the important point for this purpose is that until the interest is actually paid, there is no federal income tax deduction.

6.2.1.2 Taxation of Gain on Disposition

The taxation of the gain on the residence's ultimate disposition is a much more complicated issue due to the interaction of two very different tax rules. The first rule excludes from taxation gain on the sale of a taxpayer's "principal residence" to

[2] 24 C.F.R. § 206.33 (2007).

[3] 12 U.S.C. § 1715z–20(d)(2)(B) (2000).

[4] See I.R.C. § 163(h)(3)(A), (4)(A)(i)(I).

[5] I.R.C. § 163(h)(3)(C)(ii).

the extent that this gain does not exceed $250,000,[6] or $500,000 if the selling homeowners are a married couple.[7] Any gain realized beyond this parameter – which is not adjusted for inflation nor modified for regional differences in the cost of housing – is taxable, generally at the 15% tax rate that applies to so-called "long-term capital gains."[8]

The other tax rule that applies in this context exempts from income taxation *all* gain on assets held by a decedent at the time of that person's death.[9] This rule has no limiting dollar parameter and is not restricted to principal residences. It applies with equal force to securities of any type and to real estate that is not residential in nature, such as office buildings, warehouses, and shopping centers, among others. As a consequence, this rule provides a powerful economic incentive for people with appreciated assets to hold such assets until they die. There is to be sure, a countervailing consideration for people who might be subject to the federal estate tax, but that tax has such a large exemption – currently $2 million per person,[10] exclusive of gifts to charity[11] or to a surviving spouse[12] – that only about half of 1% of decedents faces any liability from this levy (Friedman and Aron-Dine 2006).

6.2.1.3 Economic Impact of Tax Rules for Gain on Disposition

For older homeowners contemplating a reverse mortgage, the tax law provides an economic incentive to sell the property while they are alive and residing in the residence so they can claim the residential sale exclusion, but also a counter-incentive to retain the property until they die so that the date-of-death exemption rule will apply. Recall that a reverse mortgage must be paid when the homeowner no longer lives in the residence. If that occasion arises because the homeowner has died, the date-of-death exemption rule will shelter all of the gain from income taxation. But if the homeowner has been compelled to vacate the premises for some other reason, the reverse mortgage becomes due and payable at that point. In the typical situation, the older homeowner will need to sell the residence at that time, thereby triggering the residential sale exclusion rule. And if the gain on this residence exceeds the applicable threshold ($250,000 for singles, $500,000 for married couples), some of this gain will be subject to income tax, even though it would have been completely exempted had the older homeowner been able to keep the property until she died. In effect, the reverse mortgage has the potential of exposing the gain on the disposition of a homeowner's residence to income tax when that gain might otherwise have escaped such taxation.

---

[6] I.R.C. § 121(b)(1).

[7] I.R.C. § 121(b)(2)(A).

[8] I.R.C. § 1(h)(1)(C).

[9] I.R.C. § 1014(a)(1).

[10] I.R.C. § 2010(c).

[11] I.R.C. § 2055(a)(2).

[12] I.R.C. § 2056(a).

To be sure, an older homeowner might have forfeited the economic advantage of holding property until death even if she had not obtained a reverse mortgage. That is, she may have needed to sell her residence to pay long-term care expenses in an assisted living facility or a nursing home. But other means of securing the required funds, such as selling financial assets, might have forestalled the sale of her residence. A reverse mortgage, however, forces such a sale in almost all cases, because this debt becomes due and payable as soon as the older homeowner moves out of the residence. As a result, a reverse mortgage has the effect of substantially increasing the likelihood of an economically disadvantageous pre-death disposition of the principal residence.

In any case, older homeowners with unrealized gains on their principal residence that do not exceed the residential gain exclusion's parameters are economically indifferent to this issue and may secure a reverse mortgage without further consideration. Even then, however, price appreciation that occurs after the reverse mortgage is first obtained might create a tax exposure where none existed previously And persons whose gains already exceed the relevant parameters face a serious economic disincentive to obtaining a reverse mortgage. This reality becomes more salient when one realizes that the median sales price of an existing home in the United States is $223,000 according to the most recent survey of the National Association of Realtors.[13] Even this number, moreover, masks significant regional differences; e.g., in the Western region, the median sales price is $349,500 in that same survey (see Note 13). The income tax applies, of course, only to the *gain* realized on a home sale, but with sales prices in this range, gains in excess of the residential sale exclusion are not uncommon. Thus, the tax law creates a potentially significant – and surely unintended – economic disincentive for some older homeowners who want to access the equity in their appreciated residences without selling them. The extent of this economic disincentive, moreover, is exacerbated by the essential unpredictability of many pertinent life events, such as the death of a spouse or the need to move to a medical facility.

6.2.1.4 Extended Example

To illustrate the effect of these tax rules, assume that Sharon plans to stay in her residence until she passes away. Her home cost $50,000 when she purchased it 35 years ago, but it can be sold for $500,000 today. At least three scenarios present themselves:

A. Sharon does not secure a reverse mortgage. When she dies, the entire gain of $450,000 (sale proceeds of $500,000–$50,000 cost) escapes income taxation.
B. Sharon secures a reverse mortgage but lives in the residence until her death. The same tax consequences apply as in Scenario *A*.

[13] See http://www.realtor.org/Research.nsf/files/singlefamilyreport.pdf/$FILE/singlefamilyreport.pdf (last accessed 5 July 2007).

C. Sharon secures a reverse mortgage and after several years of receiving monthly payments, she moves into an assisted living facility, a nursing home, or perhaps her adult daughter's home. At that point, the reverse mortgage becomes due. Theoretically, if Sharon can pay off the mortgage using other assets, she can continue to own the residence until her death and thereby shelter her entire gain from income tax. But the reality is that most people who obtain a reverse mortgage do so precisely because they have few other liquid assets available. As a consequence, moving out of a home that has been the subject of a reverse mortgage generally triggers a sale of that property to pay off that debt.

Such a sale will cause the recognition of $450,000 of gain. The maximum tax exclusion on the sale of her residence is $250,000,[14] leaving the remaining gain of $200,000 (total gain of $450,000 – residential exclusion of $250,000) subject to income tax at a 15% rate. As a result, Sharon's decision to obtain a reverse mortgage cost her $30,000 of federal income tax (gain of $200,000 × 15%) – plus whatever state income tax might apply – that she could have avoided by not opting for the reverse mortgage. Thus, the tax law creates a significant economic disincentive for many older homeowners to use reverse mortgages as a means of supplementing their post-retirement income.

6.2.2 *When to Begin Taking Social Security Benefits*

A central dilemma in elder law is determining when is the most opportune moment to commence the receipt of retirement benefits under the Social Security system, an inquiry that is especially well-suited to a law and economics approach. To be sure, some people might commence such benefits out of a fear that the Social Security system will soon run out of money (see Kaplan 2005a, p. 1; see generally, Kaplan 1995, pp. 198–199), but most people approach this question with the objective of determining the most economically advantageous course of action for their particular circumstances. This section explores the components of the economic analysis that is part of that determination.

6.2.2.1 Early Retirement Option

The U.S. Social Security system authorizes retirement benefits to begin as early as age 62 at the election of the prospective beneficiary.[15] But any recipient who commences these benefits prior to reaching his or her full "retirement age" faces a

[14] I.R.C. § 121(b)(1).

[15] 42 U.S.C. § 402(a)(2)(2000).

reduction in the amount of those benefits to take account of the early starting date.[16] For most of Social Security's existence, a person's full "retirement age" was 65 years, but a 1983 reform of the system raised that age gradually to age 67, depending upon the year of a person's birth.[17] We are currently in the middle of that phase-in period such that persons born between 1943 and 1954 have a full "retirement age" of 66 years.[18] If such a person chooses to begin receiving Social Security benefits at age 62, the "early retirement" formula reduces her monthly benefit by 25% (see Frolik and Kaplan 2006, p. 290 (explaining the computation)). So, if Hannah would otherwise be eligible for a monthly Social Security retirement benefit of $1,000 at age 66, she will receive only $750 should she begin taking her benefits at age 62. This reduced payment, moreover, will not "return" to $1,000 when Hannah reaches full "retirement age." That is, the early retirement reduction is *permanent*, a fact that may have long-term economic consequences to Hannah.

6.2.2.2 Delayed Retirement Option

On the other hand, if a person chooses to defer receipt of Social Security retirement benefits past her full "retirement age," a "delayed retirement" bonus is added to the payments that she receives.[19] The amount of these "delayed retirement credits" varies by a person's year of birth but at this point, it is 8% per year for each year after that person's full "retirement age."[20] For example, if Hannah in the previous example delayed her retirement benefits until she reached age 68, she would be entitled to two years of "delayed retirement credits" or a 16% addition (2 years × 8%), producing a monthly Social Security payment of $1,160 ("full" benefit of $1,000 + 16%, or $160). There is a maximum age, however, for earning "delayed retirement credits" of 70 years, after which any further postponement of benefits does not yield additional funds.[21]

6.2.2.3 Impact of Inflation

All Social Security benefits are adjusted annually for the national cost-of-living increases,[22] and these percentage increases are independent of whether the benefits received are early, on-time, or delayed retirement benefits. That is, the annual percentage "cost-of-living" adjustment (COLA) would apply either to Hannah's

---

[16] 42 U.S.C. § 402(q).
[17] 42 U.S.C. 416(*l*)(1).
[18] 42 U.S.C. 416(*l*)(1).
[19] 42 U.S.C. § 402(w).
[20] 42 U.S.C. § 402(w)(6)(D).
[21] 42 U.S.C. § 402(*l*)(2)(A).
[22] See 42 U.S.C. § 415(i).

"early retirement" benefit of $750, her "full retirement" age benefit of $1,000, or her "delayed retirement" benefit of $1,160, in each case starting from the year in which she reached age 62. Accordingly, from an economic perspective, anticipated inflation need not enter into the decision-making process with respect to starting the receipt of Social Security retirement benefits, because the COLA percentages from age 62 forward are added to whichever retirement benefit that Hannah selects.

6.2.2.4 Longevity Considerations

From the government's perspective, the "early retirement" penalty and the "delayed retirement credits" are calculated to be actuarially neutral. That is, the government is economically indifferent as to when Social Security beneficiaries elect to commence receipt of their retirement benefits. Individual elders, of course, are not so indifferent and that is why this issue is so important in elder law. "Early" benefits are smaller in amount but will be received for more years, *ceteris paribus*, while "delayed" benefits are larger in amount but will be received for fewer years, once again *ceteris paribus*. The question then becomes what the person estimates will be his or her life expectancy. To put this issue in the starkest terms, getting less now in exchange for more later makes sense only if there is, in fact, a "later." Thus, the question inevitably turns at the outset on such factors as one's personal medical history, including that of one's natural parents, if known, as modified by subsequent medical developments. That is, if a parent died at a relatively early age due to an ailment that is now treatable, that medical reality must be considered as well.

6.2.2.5 Multiple Decision Points

The elder's decision is actually more economically complicated, because commencing retirement benefits is not limited to a few specific ages. The Social Security law's formulae for "early retirement" penalties and "delayed retirement credits" are actually calibrated in terms of *months* prior to, or following, a person's full "retirement age."[23] Thus, a person might start receiving "early" benefits at age 62 years and three months, or 63 years and five months, or 64 years and eight months, and so forth. Similarly, the 8% annual "credits" are adjusted so that a person receives a bonus for delaying the start of retirement benefits for each month of such delay, not simply entire years. As a consequence, there are no fewer than 96 possible decision points. That is, the eight years between the "early retirement age" of 62 years and the maximum earning of "delayed retirement credits" of age 70 translates into 96 starting points (8 years × 12 months). Moreover, a decision to not

[23] See 42 U.S.C. § 402(q), (w).

start benefits in any given month does not preclude commencement of benefits in some subsequent month.

6.2.2.6 Implications for the Spouse

The Social Security statute provides that the spouse of a retired worker is entitled to a "spousal benefit" of half of what the worker would receive when that person would reach "full retirement age."[24] Thus, if Hannah in the previous example were entitled to receive $1,000 when she turns 66 years old, her husband Sol would be eligible for $500 at his full "retirement age" and lesser amounts should Sol begin taking spousal benefits prior to that date (see Frolik and Kaplan 2006, pp. 302–303). Sol would receive this benefit, however, only if the worker's benefit that is based on his own earnings record were less than $500. That is, Social Security pays the spouse of a retiree the higher of that person's own worker's benefit (if any) and 50% of the spouse's benefit.

But the significant point for this analysis is that a spousal benefit may be claimed only if the worker spouse is *also* receiving Social Security retirement benefits.[25] In other words, Hannah's decision whether to start receiving Social Security benefits determines whether Sol can receive benefits as the spouse of a retiree. Accordingly, if Hannah decides to *not* start receiving her benefits, Sol cannot receive spousal benefits based on Hannah's work record. Thus, married couples have this additional economic consideration to incorporate into their decision-making process. In this context, it should be noted that the federal "Defense of Marriage Act" defines marriage exclusively as between "one man and one woman" for purposes of all federal statutes,[26] and the Social Security law is a federal statute.

The Social Security law also provides that a surviving spouse (and in some cases a surviving divorced spouse) succeeds to the deceased person's actual Social Security benefit.[27] For example, assume that Hannah in the previous example was married to Sol and that Sol survived Hannah's death. Depending on the amount of Sol's Social Security benefit based on his own work record, he might be entitled to a surviving spouse benefit from Hannah if that amount exceeds his own worker's retirement benefit. Hannah's decision to elect "early retirement" or "delayed retirement," therefore, might impact how much money Sol receives after Hannah's death. In other words, the decision about when to commence receipt of Social Security benefits must consider not only the individual's projected life expectancy, but also that of a potential surviving spouse. And this need to include the expected present value of these survivors' benefits in the benefit commencement decision will be especially acute if the elder's spouse is younger or in better health than the elder in question.

---

[24] 42 U.S.C. 402(b)(2), (c)(3).

[25] Soc. Sec. Rul. 64–52, C.B. 1960–65 at 3.

[26] 1 U.S.C. § 7 (2000).

[27] 42 U.S.C. § 402(e)(2)(A), (f)(3)(A).

6.2.2.7 Post-Retirement Employment

The economics of this decision is necessarily bounded by a separate but related decision about whether the elder plans to engage in compensated employment after commencing benefits and to what extent. That is, does the elder anticipate supplementing his Social Security benefit with part-time or full-time but perhaps less remunerative employment, including self-employment? At the outset, it might seem incongruous for someone receiving Social Security retirement benefits to continue working, but both economic and non-economic factors may be at play.

Availability of Health Insurance

The most significant economic factor is probably the cost of health care. A beneficiary without health insurance could find himself quickly overwhelmed economically if a major illness or accident befell him in retirement. And most Americans receive health insurance through their employer, at least if they are working on a full-time basis.[28] While a retiree might secure health insurance on his own following his departure from the workforce, the cost of individually issued health insurance can be extremely high. That is, when the new retiree was part of an employment-based group, his insurance premiums were being subsidized economically by others in that same group. But a 62-year old retiree seeking health insurance coverage on his own will lack that intra-group subsidy. For the same reason, health insurance might be unavailable regardless of cost. That is, when this person was part of an employment-based group, acceptance was guaranteed as the insurer was obligated to accept all current employees (see Kaplan 2005b, p. 541). When the retiree is on his own, however, insurability will be determined entirely by his personal medical profile, and in many cases, that will translate into no health insurance at all.

This situation, of course, was precisely why the federal government's health program for older Americans, Medicare, was created (see generally, Marmor 2000). But Medicare is generally not available until the prospective beneficiary is 65 years old.[29] That is, unlike Social Security's "early retirement age" option which allows a person to receive benefits – albeit permanently reduced benefits – starting at age 62, Medicare has no comparable "early retirement" option available. As a result, a person who is planning to start Social Security benefits prior to reaching age 65 must consider the economic cost and availability of private health insurance or should expect to work for an employer that will cover this person under that employer's group health insurance policy. Indeed, even a person who is 65 years

---

[28] See Fronstin 2007, p. 4 (62.7% of non-elderly have employment-based health insurance, comprising 90% of those with non-public coverage), available at http://www.ebri.org/pdf/briefspdf/EBRI_IB_05–20074.pdf (last accessed 10 July 2007).

[29] 42 U.S.C. § 1395c (2000).

old might need the automatic acceptance of an employer's group health insurance coverage if his spouse is not yet 65 years old and is uninsurable on her own due to pre-existing medical conditions, a not uncommon situation for persons in this age category. Thus, the need for health insurance is a major economic consideration in the commencement-of-benefits decision.

Possible Negative Impact on Retirement Benefits

Any Social Security recipient who has not yet attained the applicable full "retirement age" faces an onerous "retirement earnings test" that substantially reduces the economic rewards from working.[30] Basically, any wages or self-employment income above an annually adjusted threshold lowers the recipient's Social Security benefits by $1 for every $2 above that threshold.[31] For example, assume that Stacey is 63 years old and earns $21,560 from part-time employment in 2008. The applicable threshold for this year is $13,560,[32] so Stacey has $8,000 of "excess" earnings (earnings of $21,560 – threshold of $13,560). This "excess" will then reduce her Social Security benefits by half of this amount – namely, $4,000. So, if Stacey's annual Social Security benefit would have been $9,300 but for this "retirement earnings test," her benefit will instead be only $5,300 (original benefit of $9,300 – reduction of $4,000).

This provision has the same economic impact as a 50% marginal tax rate on the affected earnings. Those earnings, moreover, are subject to a federal income tax on income generally of at least 15%[33] in addition to Social Security's effective 15.3% payroll tax on wages[34] and self-employment income,[35] a combined effective marginal tax rate of over 80% and possibly even more, depending upon a retiree's other sources of income.[36] State income taxes would raise this effective marginal tax rate still higher.

On the other hand, the "early retirement" penalty that Stacey incurred by electing to receive Social Security retirement benefits before she reached full "retirement age" will be recalculated when she reaches that age to reflect the loss of benefits she suffered this year. In effect, she will be treated as retiring some number of months later than she actually retired. But that adjustment is small consolation in

---

[30] 42 U.S.C. § 403(b)(1), (f).

[31] 42 U.S.C. 403(f)(3). During the year in which a person reaches full "retirement age," this test is applied on a monthly basis, and benefits are reduced by $1 for every $3 above the applicable threshold.

[32] Social Security Administration, Table of Automatic Increases, available at http://www.ssa.gov/OACT/COLA/autoAdj.html (last accessed 17 Oct. 2007).

[33] See I.R.C. § 1(c).

[34] See I.R.C. §§ 3101(a), (b)(6), 3111(a), (b)(6).

[35] See I.R.C. § 1401(a), (b).

[36] Additional income of $19,940 in 2008 would push this taxpayer into the 25% federal income tax bracket, raising that person's effective marginal tax rate to 90.3%.

the current year, and its salutary effect is entirely contingent on her future longevity. In brief, the operation of the "retirement earnings test" acts as a major economic disincentive to take Social Security benefits and engage in any remunerative activity beyond a very low level until the claimant reaches his or her "full retirement age."

Possible Positive Impact on Retirement Benefits

Once a person reaches "full retirement age," the "retirement earnings test" described above no longer applies,[37] and additional income from employment has no direct impact on the amount of that person's Social Security benefits. In fact, employment from that point on might actually increase a person's Social Security benefits, depending upon that person's prior work history. That is, Social Security's retirement benefit is based on a person's "average indexed monthly earnings" (AIME), a construct that is based on a person's 35 highest years of annual earnings.[38] If some of the 35 years that were used to calculate a person's Social Security retirement benefit had very low or no earnings, higher earnings in a later year would substitute for one of those low-or-no earnings years and would thereby raise her AIME, in turn raising her Social Security retirement benefit. This possibility is particularly likely if the elder had periods of low or no earnings due to child-rearing responsibilities, care of an older relative, extended higher education, or the like.

The possibility of this economic enhancement, however, is subject to two caveats. First, only earnings that are subject to the Social Security payroll tax are counted for this purpose, and there is an annual cap on such earnings. In 2008, that cap is $102,000,[39] which means that any earnings received that year in excess of this amount are simply ignored when recomputing a person's AIME. Second, the relationship of Social Security benefits to a person's AIME is not isomorphic. That is, the Social Security statute employs a deliberately bottom-weighted formula[40] that has the economic effect of moderating the impact of higher earnings on Social Security benefits. As a result, if Jack's AIME is twice that of Jill's, Jack's Social Security benefit will be higher than Jill's but not twice as high (see Frolik and Kaplan, 2006, pp. 294–295 illustrating computation of Social Security monthly benefit). This effect is especially pronounced as one's AIME gets above the first tier (or "bend point" in Social Security's peculiar argot), which in 2008 was $711 per month.[41] At that point, further increases in a person's AIME will increase that person's Social Security benefit but by increasingly smaller amounts.

[37] 42 U.S.C. § 403(f)(3).

[38] 42 U.S.C. § 415(b)(2)(A)(i), (B)(iii); see also Sacks 2006 p. 233; Streng and Davis 2001 paragraph 24.04[4][a], p. 24–18.

[39] Social Security Administration, Table of Automatic Increases, available at http://www.ssa.gov/OACT/COLA/autoAdj.html (last accessed 17 Oct. 2007).

[40] 42 U.S.C. § 415(a)(1)(A) (2000).

[41] Social Security Administration, Table of Automatic Increases, available at http://www.ssa.gov/OACT/COLA/autoAdj.html (last accessed 17 Oct. 2007).

6.2.2.8 Alternative Sources of Funds

A further economic consideration in the decision about when to start receiving Social Security retirement benefits is what other sources of funding are available if an elder chooses not to start those benefits. That is, what would the person live on in the absence of Social Security benefits?

One possibility, of course, is employer-provided pensions and/or retirement-oriented savings arrangements such as Individual Retirement Accounts (IRA), deferred salary arrangements under sections 401(k), 403(b), and 457, and Roth-type retirement accounts. Some employer-based pensions, however, do not pay benefits before some specified age or do so only with fairly heavy "early" retirement reductions. In any case, pension plan payments are almost always fully taxable, as are withdrawals from most IRAs and deferred salary arrangements.[42] Only Roth-type retirement accounts permit tax-free withdrawals if the accounts have been open at least five years.[43] Even those withdrawals, however, lose the benefit of further tax-free growth once those funds are taken out of their respective accounts, a major economic drawback to such withdrawals.

Social Security benefits, in contrast, are taxable only in part, that part depending upon a person's income from all sources,[44] including interest on municipal bonds,[45] which is otherwise free of federal income tax.[46] The exact formula does not warrant explication here (Frolik and Kaplan 2006, pp. 317–320 (illustrating the computations involved)), but a few parameters merit mention:

- If an individual recipient has annual income from all sources of less than $25,000, or a married couple has annual income of less than $32,000, Social Security benefits are not taxable at all[47]
- If an individual's income is between $25,000 and $34,000 (or a couple's income is between $32,000 and $44,000), a portion of their benefits, perhaps as much as 50% of those benefits, is taxable, with the exact percentage rising proportionately within the specified range[48]
- If an individual's income exceeds $34,000 (or a couple's income exceeds $44,000), a higher percentage of those benefits, but never more than 85%, is subject to income tax.[49]

In any case, the bottom line is that *only a portion* of Social Security benefits is subject to tax, in contrast to most retirement account withdrawals.

---

[42] I.R.C. §§ 61(a)(11), 402(a).

[43] I.R.C. § 408A(d)(1), (2)(B).

[44] I.R.C. § 86(b)(1)(A)(i), (2).

[45] I.R.C. § 86(b)(2)(B).

[46] I.R.C. § 103(a).

[47] I.R.C. § 86(b)(1)(A), (c)(1)(A), (B).

[48] I.R.C. § 86(a)(1), (2), (c)(2).

[49] I.R.C. § 86(a)(2), (c)(2)(A), (B).

Another possible source of retirement funding is liquidation of non-retirement assets. Most of those assets, especially stocks and mutual funds as well as investment real estate, would qualify for a preferential "long-term capital gains" tax rate of 15% or even less in some circumstances.[50] In contrast, to the extent that a person's Social Security benefits are taxable at all, those benefits would be taxed at "ordinary income" rates, which can go as high as 35%.[51] And as noted earlier in this chapter, if one of those assets is the elder's principal residence, the first $250,000 (or $500,000 if married) of gain realized from the sale of that residence would be received free of income tax.[52] In other words, many older persons might experience lower tax rates by disposing of their non-retirement assets while postponing the receipt of Social Security benefits. On the other hand, electing to take Social Security benefits might enable the elder to keep her appreciated assets until she dies, in which case the entire accumulated gains would be excused from incurring any income tax, via the date-of-death rule discussed previously.[53] Obviously, any given elder's entire economic situation must be evaluated individually.

6.2.2.9 Investment Strategy

Some elders might start receiving Social Security benefits, even if they do *not* need these funds to live on, in order to take the money and invest it elsewhere for a higher anticipated rate of return. The economics of Social Security's "early retirement" penalty and "delayed retirement credits," however, suggests that the break-even point in this strategy is about 6.7% to 8%, depending upon whether the elder in question has reached "full retirement age" (see Frolik and Kaplan, 2006, pp. 290, 292). This result, moreover, must be obtained on at least a partially after-tax basis, depending upon which tier of taxation of Social Security benefits applies to the specific elder. Further, the relevant increase in Social Security benefits is mandated by federal statute,[54] so the economic comparison must be to investments with zero risk. As a result, if the elder plans to invest his Social Security benefits in bank certificates of deposit, money market funds, or US Treasury obligations, it is extremely unlikely that the yield on those financial instruments – all of which are fully taxable as "ordinary income" – will exceed the available increase in Social Security benefits. Only if that person contemplates much riskier investments will the strategy of take-the-benefits-and-invest-elsewhere work out economically.

---

[50] I.R.C.. §§ 1(h)(1)(B),(C), 1221(a) (definition of a "capital asset").

[51] I.R.C. § 1(a)-(d).

[52] I.R.C. § 121(b).

[53] I.R.C. § 1014(a)(1).

[54] See 42 U.S.C. § 402(q), (w).

6.3 Conclusion

This chapter has shown how the law and economics approach is critical to understanding two of the major questions in elder law that older people confront. This approach has considerable utility beyond these two issues, of course; some leading proponents, in fact, apparently sense no bounds whatsoever in its applicability. It is not necessary to endorse that perspective, however, to appreciate the layered and especially rich contours that law and economics can bring to elder law, particularly in assessing whether certain statutes accomplish their intended objectives or simply confound their presumptive beneficiaries and frustrate sensible public policy.

This chapter's consideration of reverse mortgages and Social Security's benefit options suggests that radical reform and simplification are needed for some of elder law's core regimes. In the absence of such revisions, older people must struggle with multi-faceted statutes that directly impact their well-being but would strain the decision-making capabilities of even much younger persons. Do such complicated schemes provide individually tailored solutions or simply spawn more frustration?[55] According to Professor Ulen, the "rational choice theory" that undergirds so much of law and economics posits that: "people are rationally self-interested decision makers; *they are capable of computing the costs and benefits of the various alternatives open to them*; and they seek to choose that alternative that is likely to give them the greatest happiness" (Ulen 1996, p. 109 (emphasis added)). But how realistic is this proposition? This chapter has shown that in at least two key elder law contexts, the theoretical foundation of law and economics that everyone is economically driven and well-informed about the costs and consequences of the alternatives available may be more aspirational than descriptive. No one is particularly well served by that state of affairs.

References

Bittker BI et al (2002) Federal income taxation of individuals, 3rd edn. Warren Gorham & Lamont, Valhalla, NY

Cooter R, Ulen T (2000) Law and economics, 3rd edn. Addison-Wesley, New York

Friedman J, Aron-Dine A (2006) The state of the estate tax as of 2006. http://www.cbpp.org/5–31–06tax2.pdf. Accessed 5 July 2007

Frolik LA, Brown MC (2000) Advising the elderly or disabled client, 2nd edn. Warren Gorham & Lamont, Valhalla, NY

Frolik LA, Kaplan RL (2006), Elder law in a nutshell, 4th edn. Thomson/West, St.Paul, MN

Fronstin P (2007) Sources of health insurance and characteristics of the uninsured: updated analysis of the March 2006 current population survey, Employee Benefit Res. Inst. Issue Brief No. 307

---

[55] Cf. Kaplan (2005c), Analyzing the range of choices presented by Medicare's enacted drug benefit plan and the impact – often irrevocable – of such choices on older people's existing mechanisms for purchasing prescription medications.

Hammond CM (1993) Reverse mortgages: a financial planning device for the elderly. Elder Law J 1:75

Heller PS (2006) Is Asia prepared for an aging population? International Monetary Fund Working Paper WP/06/272

Kaplan RL (1995) Top ten myths of social security. Elder Law J 3:191

Kaplan RL (2005a), The security of social security benefits and the president's proposal, Elder Law Rep 1

Kaplan RL (2005b) Who's afraid of personal responsibility? Health savings accounts and the future of American health care, McGeorge L Rev 36:535

Kaplan RL (2005c), The medicare drug benefit: a prescription for confusion. NAELA [Nat'l Acad. of Elder L. Att'ys] J. 1:167

Marmor TR (2000) The politics of medicare, 2nd edn. A. de Gruyter, New York

Posner RA (1995) Aging and old age. University of Chicago Press, Chicago

Posner RA (2003) Economic analysis of law, 6th edn. Aspen, Gaithersburg

Regan JJ et al (2007) Tax, estate and financial planning for the elderly. Lexis Nexis, Newark, NJ

Reilly J (1997) Reverse mortgages: backing into the future. Elder Law J 5:17

Sacks AL (2006) Social Security Explained. CCH Inc., Chicago

Scholen K (1993) Retirement income on the house. NCHEC Press, Apple Valley, MN

Streng WP, Davis MR, Retirement planning: tax and financial strategies, 2nd ed. Warren Gorham & Lamont, Valhalla, NY

Ulen TS (1996) The law and economics of the elderly. Elder Law J 4:99

Chapter 7
What can Elder Law Learn from Disability Law?

D. Surtees

7.1 What is Disability Law?

The question "What is disability law" is quite different from the question "what is disability?" I think the first question (about law) is considerably easier to answer than the second (about disability). Yet the answer to the first question remains woefully incomplete until we address, in some significant measure, the disability question.

'Disability law' of course is the area of law which concerns itself with the interaction of law and people with disabilities. I shall call these people members of the disability community. In using this term I am cognizant that some would contend individuals with disabilities do not form a community. Others would point to certain groups, such as the hearing impaired, who share many common cultural attributes and say there are actually several disability communities. I am simply using the term as one way to refer to the somewhat vaguely defined group of people who share the attribute of being labeled as having a disability.

Sometimes the law affects members of the disability community by extending certain benefits or entitlements to them.[1] Other times the law affects members of the disability community by providing protection to them.[2] In both cases it is necessary for the law to determine community membership in order to function. Where

---

[1] For example, the provision of a placard or tag which allows a driver to park in designated 'handicapped' parking spaces when the vehicle is driven by, or is transporting, a person with a disability, as defined by the appropriate legislation. For a critical analysis of handicapped parking regulation see: Miller and Singer (2000–2001, pp. 81–126).

[2] For example human rights legislation may prevent a landlord from discriminating against a potential tenant on the basis of disability. An example of this type legislation is *The Saskatchewan Human Rights Code*, S.S.1979, c. S-24.1, ss. 11(1) which provides "No person, directly or indirectly, alone or with another, or by the interposition of another shall, on the basis of a prohibited ground:

(a) deny to any person or class of persons occupancy of any commercial unit or any housing accommodation; or

(b) discriminate against any person or class of persons with respect to any term of occupancy of any commercial unit or any housing accommodation."

I. Doron (ed.) *Theories on Law and Ageing: The Jurisprudence of Elder Law*,

the law has different purposes, the line determining who is within the community will be drawn in different places. A statute such as a human rights statute which prohibits discrimination against people with disabilities ought to define the group more expansively than a statute extending a benefit such as an entitlement to a pension. A person who is merely perceived as having a disability, where no disability in fact exists for that person would be properly included in the human rights definition and properly excluded from the pension entitlement definition.

Disability law is one of several areas of law defined primarily by identification of a group within society as opposed to an area of law defined primarily by the cohesion of the substantive rules within the area – like contract law or tort law. This means that disability law is also a lens though which the impact a law has on members of this community may be evaluated. Using disability law as a lens makes it applicable to almost any area of law. This aspect of disability law asks the question "How will the development of the law in this area affect members of the disability community differently than non-members?" The answer to this question still requires some working understanding of who counts as a person with a disability. However, different people can be included or excluded from that definition depending upon the area of law being examined. Where zoning of residential housing is being considered for example, the differential impact of the law may be limited to those members of the disability community who are likely to live in housing arrangements which differ from the traditional "one family – one home" model. There is virtually no area of law or policy which cannot to some extent be looked at through this lens. Particularly important examples include employment, education, health care, the right to life and the right to die. Other examples abound.

Disability law then includes the study of the interaction of law and members of the disability community. This interaction may be direct or indirect. It can include value judgments with which our laws are imbued. It can include interactions as diverse as standards for the development of parks and motor vehicle licencing regulations. The interaction may create positive or negative impacts. The impact may arise from the extension of state protection from discriminatory action or it may arise from the provision of a benefit under the law. Finally disability law acts as a lens to help us see the differential impact any law or policy may have on a person with a disability. Understanding each of these aspects of what disability law is requires some examination of "who counts" as a person with a disability. Understanding this question raises the related question of "What counts as a disability?" Paradoxically, I believe that more can be learned by understanding *how* we have answered that question than can be learned from the answer itself.

7.2 Milestones in Understanding

Disability has always been with us. A history of disability is really a history of a society's reaction to disability. It may chronicle the way society understood or defined disability. It may describe 'treatments', sometimes voluntarily submitted to and other times forced upon people. It may try and explain changing public attitudes. A history of disability within a society is very much a history of the society itself. In this way Disability Studies acts like a mirror to inform those concerned with law and disability. Understanding how we have answered the question "What counts as a disability?" can inform legal analysts so that we can more accurately identify the impacts, both intended and unintended, of a particular law or policy upon the disability community.

Academics typically apply ways of understanding developed in one area of the law to a different area of the law. Through this approach, one can develop new and exciting insights into the area of law being studied. This is what I am attempting to do by applying my understanding of disability law to elder law. Others have viewed disability law through lenses such as Critical Race Theory and Feminism (Asch 2001 p. 391), Feminism and Communitarian Theory (Ball 2005 p. 105), Law and Economics (Stein 2003 p. 79), Critical Disability Theory (see for example, Pothier and Devlin 2006), and other lenses.

I suggest that the starting point of our modern understanding of what counts as a disability is the location of the disability within the person. This is a very individualized understanding of disability. 'A person with one leg is disabled because they are an amputee.' Understanding disability in this way provides fertile ground for certain developments to occur. Disability, understood as a deficiency, must be a negative concept. A person who could barely run and therefore displayed an extreme negative variance from the norm, might be labeled as disabled. No one would label a person who could run better than anyone else, and therefore displayed an extreme positive variance from the norm, as disabled. Disability is therefore a fault, a flaw in the person. This is more sophisticated than some ancient understandings of disability. We no longer believe that disability has "supernatural or demonological causes" (Braddock and Parish 2001, p. 17). Modern understanding of genetics and other sciences made paranormal explanations unnecessary. Yet the understanding of disability as individual flaw remained. Only our understanding of causation grew more sophisticated.

Institutionalization of some members of the disability community has been a typical reaction in many societies (see generally Braddock and Parish 2001, p. 17). The first Canadian residential institution built for individuals with developmental disabilities was the so-called "Asylum for Idiots". It opened in Orillia, Ontario in 1859 – 18 years prior to the forming of the Canadian confederation (The Roeher Institute 1996, pp. 3–5). For over a century, institutions would figure prominently in many governments' response to disability. The doors of many institutions would eventually close for the last time in the period between the 1970's and the 1990's.

While institutionalization served to remove those labeled disabled from sight,[3] care for the individual, and treatment of the disability, did occur within many institutions. The process of de-institutionalization particularly for those with developmental disabilities was accelerated in North America around the 1970s (see for example

[3] "[The English] Parliament enacted a law in 1714 authorizing confinement, but not treatment for the "furiously mad" (Stein 2003, p. 25).

Feleger and Boyd 1979 p. 717). In Canada, each of the provinces has been engaged in a decades long process of deinstitutionalization of individuals in psychiatric hospitals (Sealy and Whitehead 2004, pp. 249–257). During the fifteen year period 1964–1979, the number of psychiatric hospital beds in Canada decreased from 4 per 1,000 people to 1 per 1,000 people (Sealy and Whitehead 2004, p. 250).

Viewing disability as a characteristic of an individual leads logically to 'treatment' – that is the correction of the defect. This is not the same thing as rehabilitation, but the two concepts are related. "[T]he goal of rehabilitation is to increase an individual's range of skills and abilities to function more independently and become a productive member of society" (The Roeher Institute 1996, p. 15). In both cases the individual is seen as a person with a defect.

The understanding of disability as an individual defect, and the grouping of significant numbers of individuals with those defects in institutions, supported a 'professionalization' of treatment. The Binet-Simon scales for measuring intelligence were published in 1905. This provided an accepted and seemingly objective method of quantifying intelligence. There was an appearance of scientific precision to the process. Intelligence was seen as hereditary. As the eugenics movement gained influence, the involuntary sterilization of developmentally disabled women became common place. In hindsight this seems to have been a logical outgrowth of the desire to 'cure' disability. No doubt, it was made more likely by the presence of institutions and the belief in what has been described as the medical model of disability.

As our understanding of the medical sciences grew, a logical result of this view of disability was the development of the 'medical model' of disability. The placement of many individuals, including children, in state-run institutions would certainly facilitate the professionalization of treatment and attempted cures. Creating institutions, with medical staff on-site, permanently or itinerantly, would make experimentation easier. The more disability came to be understood in medical terms, the more non-medical personnel were excluded from the discussion. And so the medical model became the pre-dominate model by which we answered the question "What counts as a disability?" Although we struggle with specific definitions, we share an understanding that to be a person with a disability requires a medical diagnosis of some affliction located within the individual.

Although no longer seen as the primary model for understanding disability[4] elements of the medical model served, and continue to serve a valuable role. The model provides an efficient and apparently objective method of gate keeping. This is important for entitlement programs. Entitlement programs are simply programs for which one must have a disability to gain entry. Whether it involves eligibility to attend a special summer camp, the right to park in certain parking spots or the right to special needs equipment, someone has to be able to determine if a particular

[4] "In the late 1960s, a fundamental transformation occurred in [American]federal disability policy that rejected a primarily medical/clinical model of disability and substantiated a socio-political or minority group model" (Scotch 2000 pp. 213, 214).

person qualifies. However this is phrased, it really comes down to defining disability for the purposes of the program, and then determining if a particular individual is disabled within that definition. Relying on a medical model to do this has several advantages for the service provider. It appears neutral and objective. That is, once the service provider has crafted the definition of disability, all it needs to do is require the applicant to obtain a medical certificate indicating a doctor's diagnosis. This also makes the process efficient for the service provider. Its decision making role is now very limited. It simply has to determine if the diagnosis correlates with the definition. There is no longer any need for discretion. In fact one would presume that prospective applicants who do not receive the required diagnosis would often simply withdraw from the application process thus removing the need for the service provider to determine that the diagnosis is insufficient. Where the diagnosis is sufficient the doctor could merely sign a certificate stating that the person examined meets the entitlement criteria.

This medical model process minimizes the strain on the service provider in performing its gatekeeper functions. From the point of view of a service provider, the medical model retains a certain attractiveness. In some circumstances it can make gate-keeping an efficient and simple matter over which the service provider has great control because it sets the criteria to be used, but reduced responsibility because a neutral, objective medical practitioner determines if an individual meets the stated criteria. The medical model however is flawed because it does operate from the perspective of the service provider.

In appearing objective, it masks the ambiguity of diagnosis. It masks the power and subjectivity inherent in determining what the criteria will be for deciding 'who counts'. In many cases diagnosis is ambiguous. Where a patient presents with symptoms which may or may not meet the criteria established for eligibility to a program, the medical practitioner's decision will determine whether or not their patient qualifies for the program. The patient, if not the medical model itself, can put pressure on the medical practitioner who is acting as the gatekeeper for the program.

Related to the medical model (or bio-medical model as it is also called) is the 'functional' approach to disability. The medical model seeks to identify the medical cause of disability. This leads to a diagnosis of what is wrong with the individual, and ultimately, at least in successful cases, a treatment or cure of the condition. The functional approach likewise sees the disability located within the individual. However, rather than focusing on the medical or biological cause of the disability, this approach focuses on the "functional incapacity resulting from an individual impairment" (The Roeher Institute 1996, p. 15). So whereas the medical model says a person with one leg is disabled because they are an amputee (the diagnosis) the functional approach would focus on the incapacity which results from the individual's impairment. The lack of one leg (for whatever reason) causes the individual to have greatly restricted mobility. The functional approach looks to rehabilitation to overcome the functional incapacity. So, while an amputated leg cannot be "cured" the individual, through rehabilitation, can learn to walk with the use of an artificial limb. The focus in this model remains on the individual. It simply shifts from the *cause* (in biological or medical terms) of the disability to the *effect* of the disability, which is some sort of functional limitation. Strategies can then be employed to provide the individual with the skills, aids, knowledge or ability necessary to overcome the functional limitation – or at least to minimize it. As with institutionalization, disability is hidden. It is erased from society.

In both of these models the focus is on how to change the individual, so that disability is erased, or at least out of sight. There are certainly large numbers of caring, compassionate, dedicated and talented individuals who have spent their careers trying to make the lives of people with disabilities better. Doubtless, there are many success stories. Medical treatment and rehabilitation services are necessary and important. However, the systemic focus on the individual as a broken person provides an incomplete view.

In the 1960's the legal scholar Jacobus tenBroek characterized the choice for people with disabilities who received welfare and therefore had to obey welfare case workers and rehabilitation workers, as a choice between obedience and starvation. This is how he put his criticism of the system that provided disability benefits:

> It is the agency of welfare, not the recipient, who decides what life goals are to be followed, what ambitions may be entertained, what services are appropriate, what wants are to be recognized, what needs may be budgeted, and what funds allocated to each. In short, the recipient is told *what* he wants as well as how much he is wanting (ten Broek and Matson 1966, p. 809, 831 quoted in Bagenstos 2004, p. 1, 14).

Rather than locating disability solely within the individual, people began to see disability as a "complex collection of conditions, activities and relationships, many of which are created by the social environment"(Bickenbach et al. 1999). This change marks a milestone in the development of our understanding. It helps us understand tenBroek's critique of the effect the system of providing disability benefits had on those receiving the benefits. It is simply inappropriate to develop a system to assist members of the disability community by focusing exclusively on what can be done for or to them, so that they may be cured or may rehabilitate themselves.

This social model recognizes that disability is, at least in part, a social construct. We create disability when we collectively determine how to build the environment around us. We can create a social environment which is disabling to some people. This aspect of the social model, sometimes called a 'social constructionist approach"[5] recognizes that the general ways of doing things and ways of thinking which are considered normal in a society, can create disability. A society creates institutions and attitudes. When we chose to treat people differently based on a characteristic, we can create disability. Placing individuals in an institution because they have a certain characteristic is an example of this. Creating services and constructing access to social 'goods' in a way which fails to take human variation into account, also results in the disabling of some individuals. For example a social

[5] "The objective of the social constructionist approach to disability is to uncover the subtle societal factors which interplay with personal experiences and together create, reinforce and potentially perpetuate the subordination of people with disabilities" Jones and Marks (1999, p. 3).

program which requires one to regularly appear at a designated office in a busy downtown location to fill out income verification forms, would disadvantage some individuals such as those with mobility limitations which made the trip downtown difficult for them. When access to desired opportunities is consistently limited because of the way the opportunities are offered, we can create disability (see generally Burgdorf (1997, esp. pp. 516–518).

We can also create a physical environment which is disabling to some people. We can construct buildings with stairs, narrow doorways, door knobs which are difficult to turn, and high countertops. If we construct our physical environment in a way which is disabling to some people, we have created disability. By the same token, we create disability by our social norms. We determine what we consider to be 'the norm' and therefore we determine what constitutes variation from the norm (Stein 2007, p. 75, 86). The philosopher Anita Silvers said, "If the majority of people, instead of just a few, wheeled rather than walked, graceful spiral ramps instead of jarringly angular staircases would connect lower to upper floors of buildings" (Silvers 1998, quoted in Stein 2007, p. 87, footnote 61)

The unaccommodating physical environment and majority social beliefs work together in this model, to create disabling conditions. Therefore remedial efforts ought to seek to change the environment and change social attitudes. No longer is disability seen as an individual flaw. At the same time the social model is perfectly capable of recognizing the value of medical treatments, and the importance of using adaptive technology and learning adaptive skills. However, if we are to understand disability, we must look to the social and physical environments we have created. If something must be changed to accommodate people, we can and we should change the environment which we created, and which in turn is creating disability.

This way of understanding disability is significantly different from the medical model. This difference will be reflected in policy implementation. The change became visible in international instruments beginning in the 1970's and became more pronounced in the 1980's, which the UN proclaimed the International Decade of Disabled Persons (Stein 2003, pp. 88–89).

An outgrowth of the social model is the characterization of people with disabilities as a minority group.[6] The need for a legal right to live in the world was a theme of Jacabus tenBroek's work. This so called "minority group model" or civil rights model has been used by policy makers to bring about "institutional change" (Scotch and Schriner 1997, p. 151). It is reflected in the *Canadian Charter of Rights and Freedoms*, the *Americans with Disabilities Act*, human rights legislation and in law and policy which seeks to explain and minimize disadvantages individuals experience because they are members of the disability community.[7] The civil rights model has ushered in a great deal of positive change.

---

[6] 'The minority group analysis was an outgrowth of the scholarship and political activism that helped create the social model of disablement" Bickenbach et al. (The Roeher Institute 1996, p. 11).

[7] "The minority group model…deals with disability as a stigmatized attribute of a socially defined group, and it focuses on the use of legal remedies and on defining the problems of people with disabilities as rooted in the politics of intergroup relations" (Scotch and Schriner 1997, p. 154)

Some scholars, however, suggest that it is possible to develop a model beyond the civil rights model, which would more faithfully capture the complexity of disability (Scotch and Schriner 1997, p. 154). This is in my view, a milestone in understanding that we are only just arriving at. There are differences in humans' abilities. Some of these abilities are accommodated and others are not. We construct homes with windows and electric lights because most people are able to see, provided there is adequate lighting and most people must see in order to navigate a building. We build homes with stairs because most people are able to navigate the stairs. In the first example, adaptive technology (lights and windows) is used to accommodate the majority group, although it is of little use to a minority group who cannot see at all and navigate by using a cane or a dog. In the second example adaptive technology (a ramp) is not used even though its absence disadvantages a minority group and its presence would not disadvantage the majority group. Some human variation is accommodated; some is not.

7.3 Moving beyond Civil Rights

Jerome Bickenbach and others, are articulating a new model for conceptualizing abilities they call *universalism* (Bickenbach et al. 1999; Scotch and Schriner 1997, p. 151). Rather than focus on attributes which separate the "minority group" from the rest of society, universalism attempts to articulate a more inclusive approach.

Bickenbach articulates two main concerns with the civil rights or minority group model. The first is that he says people with disabilities simply do not share common experiences in the same way people who share the same racial group do. He says there is "no unifying culture, language or set of experiences" (Bickenbach, Jones and Marks 1999, p. 14) shared by people with disabilities. The leaders of the disability movement are not representative of people with disabilities. They tend to be highly educated, white males with few medical needs and a late onset disability (Bickenbach, Jones and Marks 1999, p. 14). People with disabilities have such variation in their experiences that many would not recognize the experience of differently situated people with disabilities. In this model, the concept of a disability community melts away.

Secondly, Bickenbach urges moving beyond a civil rights paradigm because discrimination simply does not cause all the inequality faced by people with disabilities. Where discrimination does cause a hardship or inequality, a civil rights approach may be useful in ending the discrimination. However Bickenbach says many examples of inequality are "brought about by a maldistribution of power and resource" (Bickenbach, 1999, p. 1181).

While the social model of disability has fostered many great strides toward inclusion and at times equality, it is not a model one can expect will lead us to a generalized or normalized state of equality. Law is a fundamental and important force in society. Law normalizes and law marginalizes – but so do other manifestations of power. Where law regulates the interaction between individuals it may be the truest arrow in the quiver available to puncture discrimination. Some hides are so tough, however, that they require more than a single arrow to be punctured.[8]

Accepting that the civil rights model is not the appropriate model with which to create a generalized or normalized state of equality does not in itself mean that the developing theory of universalism (to use Bickenbach's term) is any better. Bickenbach, Scotch and others suggest a theory which takes account of the infinite variation of humans. In my view, the genius of this theory is that it holds the promise of rendering normalization absolute in this context. By using a theory which embraces the infinite variation of humans, we develop policy and law based on inclusion. Universal inclusion is the promise – it is the ideal. Law normalizes by defining the acceptable and by defining boundaries. This is the same process by which law marginalizes.

If we construct a building with many stairs, we define walkers as normal and marginalize wheelers. If we use a human rights approach to attack this design on the basis of its discriminatory effect on wheelers (even though this may be unintentional) we may achieve a ruling that requires building construction to not discriminate. Some combination of ramps and lifts may achieve this. The dominant group (walkers) must accommodate the minority group (wheelers). This is not a bad result – but neither is it the best result. An approach which recognizes infinite human variation has the potential to normalize the use of the building by everyone. Therefore the question may be restated.

The human rights question is "Has this individual been discriminated against by the design of this building?" The person responsible for the building must admit this accusation or defend against it. The process is adversarial. The person responsible for the building is representing walkers. Given the building design this person is presumably a walker, but it is unimportant whether they personally walk or wheel. They are representing the interests of walkers. In adversarial matters it is not uncommon for a group to seek out a representative from the other side. While this may be a useful advocacy technique it does not alter the fundamental issue which is the adversarial contest the two groups are engaged in.

The question posed by universalism is "Has universal design been used to create a building which minimizes the effect of individual variation?" The 'solution' or result of asking the question may be that a particular building is still retrofitted with ramps and lifts – but the significant difference in framing the question has important results. The human rights model can see society as a huge number of individuals with traits. Some of these traits relate to prohibited grounds of discrimination. Each protected ground of discrimination (disability to be sure, but also gender, age, race and so on) acts as sorting tool to place the individual into a majority group, or a minority group. All of these sorting exercises result in one majority group and most result in several minority groups. Curiously, gender results in the

---

[8] "What is very important to appreciate is that even if there existed a perfect regime of human rights, a system of formal law promoting and empowering people with disabilities, this is only going to be a small part of what is necessary to bring about true equality for people with disabilities" (Jones and Marks 1999, p. 4)

majority (females) being identified as the 'minority' group, and the other group (males) being identified as the majority group. Of course some would suggest that there are other minority groups such as trans-gendered individuals. The male-female example shows however that the human rights approach is really more about the power the group has, rather than its numerical strength. The human rights approach works by dividing individuals on the basis of an attribute which results in relatively weaker minority groups being aided by the power of the law. This additional power is very useful in addressing the individual matter of discrimination under examination. However by reinforcing grouping of individuals so that we remain divided into a dominant group and minority groups, the model does little to bring people together to develop policy which is more inclusive. Further this model does nothing to bring about social change in areas outside the realm of what is regulated. Only legally discriminatory actions are capable of regulation and control. Whether an action is regulated or not depends upon both the type of interaction (employment, housing and so forth) and the delineation of characteristics deemed to be prohibited grounds of discrimination.

In contrast to the human rights model, the model based on universalism ideally defines society as a single group, made up of infinitely variable individuals. Individuals will of course group together for other purposes; social, cultural and political purposes come to mind. But for the purpose of understanding ability, universalism aims to establish a view of ourselves as individuals who exist at an infinite number of points on an infinite number of continua of ability levels. In practice, universalism will inevitably still draw a line which normalizes some differences and marginalizes others. The promise is that this line will be drawn in the most inclusive way practicable. The risk is that small minded people will urge unimaginatively narrow line drawing. It is possible to subvert universalism by the argument that 'since we must draw a line somewhere, we might as well draw it where it has usually been drawn to include and normalize the powerful and to exclude and marginalize the weak'. This risk however is not a reason to reject the model. Rather it is a reason to embrace it, to discuss it, to make sure it, and its promise, are understood. Rather than using difference to reinforce separation it is time to recognize that we are united in our variation.[9] We are a continuum of humanity.

7.4 Applications to Elder Law

If disability law is the area of law which concerns itself with how the law impacts people with disabilities, then elder law must be the area of law which concerns itself with how the law impacts those who are considered elders. The law affects

[9] "The minority group model is a very effective way of developing solidarity and collective identity, but it runs the risk of reinforcing the separateness of disability people".

Tom Shakespeare "What is a Disabled Person?" in Jones and Marks (1999, p. 31).

members of this group by extending certain benefits or entitlements to them, and also by providing certain protections to them. As with disability law, a primary function of the law with respect to law which extends benefits, is gate keeping. If benefits are not intended to be universal, some criteria must be applied to determine who is 'in' and who is 'out'. At times 'age' is the criteria used. Where the law seeks to extend protection to those we consider elders, we must establish membership in the group just as when protection is extended based on disability. In this situation, as with disability law, membership may be more leniently determined. Those who appear to be within the group are generally thought to be deserving of the protection acquired through group membership along with those who are actually members of the group.

Like disability law, elder law is in part an area of law which is concerned with evaluating the effect the law has on members of an identified group. There is a substantive component to elder law. Laws dealing with retirement from the workforce for example would substantively be part of elder law. Elder law is also a lens through which virtually all areas of law can be viewed so as to determine the differential impact the law has on group members.

While there are similarities between elder law and disability law, there are two fundamental differences which in my view make universalism even more appropriate for elder law than it is for disability law. The first of these is that we have a history of trying to 'fix' disability, but not age. The second is that those separated out from the mainstream by disability typically have a history of exercising relatively less power, and experiencing relatively greater marginalization. This is not so with those separated out from the mainstream by age.

There is certainly some overlap between those who are seen as disabled and those who are seen as elders. There is also tremendous individual variation within each group. However group membership is assigned based on 'difference'- either with respect to ability or age. Membership in one or the other brings about a distinctive reaction to the difference. Where the difference which defines a person's group membership relates to disability, the societal response historically has been to begin by changing the person. Where rehabilitation is possible and successful it preempts the need for a human rights based anti-discrimination legislation. Only where rehabilitation is not successful do we enter into a conversation as to whether we must accommodate the difference. This is not prevalent with elders. We have a long history of trying to 'fix' disabled people – or encouraging them to fix themselves. When this is not successful the law, in a human rights model, is seen as the best way of determining if we must accommodate and if so, how we must accommodate. We do not have a history of trying to fix elders before determining if we must accommodate. Some may consider elders 'unfixable'; age simply does not yield to rehabilitation efforts. Most, I would suggest, don't try and 'fix' elders because we don't see them as broken. Of course many people use plastic surgery, cosmetic injections and other techniques to try and deny the appearance of aging. Typically however, people do this for and to themselves. Even the most vain do not attempt to overcome the appearance of aging in others. Therefore when the factor which defines the difference is age, as opposed to ability level, our history is one of moving more quickly to determine whether or not to accommodate. We have as a society been more accepting of elders as elders than we have of the disabled as disabled.

This different history helps explain why the civil rights model was seen as such a step forward for people with disabilities. Certainly a model which focuses on discrimination or the need for others to change to accommodate you is a giant step forward from the alternative of being the one who must be fixed so that the need for others to accommodate disappears or diminishes. The civil rights model was an improvement, and a move closer to the universalism model for people with disabilities. It would not be much of a move forward for elders. Certainly there are individual elders and elder law issues which can and do receive helpful results from the civil rights model. As a group, however, elders generally do not have to take steps in order to move the conversation towards the need to accommodate. That is where the conversation starts.

When the factor which defines our difference is age as opposed to ability level, another distinction becomes apparent. The 'power history' of the individuals who populate each of these two groups is different. Of course within each group there will be individuals who have lived lives characterized by being marginalized and relatively powerless. There will also be individuals within each group who in most aspects of their lives are considered mainstream and individually powerful as well. Members of each group face stereotypes and discrimination which create barriers and harm. It is my contention, however, that on average when individuals are divided by ability level those who are seen as congenitally 'disabled' will live lives which will be far more likely to be presumed to be relatively powerless and marginalized. In contrast, individuals who are seen as having acquired a disability, especially if this happened in adulthood, are more likely to be seen as sharing a power history with elders. When individuals are divided by age, those considered elders are far more likely to be presumed to have lived lives characterized by having relatively greater power. That is to say that at least up to the point when an individual in placed in the group labeled 'elders', they will have, on average lived a life characterized as relatively more powerful. Even if seen as currently powerless, elders, and those who are seen as acquiring a disability as an adult, will generally be taken as at least once having been powerful. In this respect the presumptions faced by those seen as having a congenital disability (or presumably having acquired a disability as a child) will differ from the presumptions faced by the other two groups.

In reality, members of each group probably represent the full spectrum of marginalization. Some will live quite marginalized lives, others will not. On average however, members of the elder group will have more experience with having power and being part of the dominant group than will members of the disability group. This, I believe, makes a civil rights approach both a less necessary and a less desirable approach to elder law. Universalism offers a more natural fit with elders. It offers greater hope of developing law and policy which will be effective in achieving systemic reforms in areas like housing, employment, health care, which will promote optimal independence and participation in society for people regardless of age.

Madam Justice L'Heureux-Dubé said:

> Justice and equality play an important role in the process of making the law relevant to the aging generation. This does not mean special rights for aging people but, as the United Nations' slogan indicates, rights that ensure "a society for all ages" ("Special Note" in Sodden 2005)

Universalism as a model to understand elder law carries with it the hope that all of us can be united in designing programs and policy which include us all, wherever we currently find ourselves on time's continuum. Age is a circumstance which, when combined with some life situations, can impact upon a person's vulnerability. Age should not be ignored. The civil rights model, however, risks using age to divide us. It uses age to determine if a certain person is an 'elder' or not, in the same way as it asks if a person is 'disabled' or not. It is divisive. When we are divided some are marginalized. A continuum of age should be used for inclusion, not to divide us. Universalism holds this promise.

References

Asch A (2001) Critical race theory, feminism and disability: Reflections on social justice and personal identity. Ohio St L J 62:391

Bagenstos SR (2004) The future of disability law. Yale L J 114

Ball CA (2005) Looking for theory in all the right places: Feminist and communitarian elements of disability discrimination law Ohio St L J 66:105

Bickenbach J et al (1999) Models of disablement, universalism and the international classification of impairments, disabilities and handicaps. Social Sci Med 48:1173

Braddock D, Parish S (2001) An institutional history of disability. In: Albrecht G, Seelman K, Bury M (eds) Handbook of disability studies. Sage, Thousand Oaks, CA

Burgdorf RL (1997) "Substantially Limited" protection from disability discrimination: The special treatment model and misconstruction of the definition of disability. Villanova L Rev 42:409

Feleger D, Boyd P (1973) Anti-institutionalization: The promise of the Pennhurst case. Stan L R 31:717

Jones M, Marks LAB (1999) Law and the social construction of disability. In Jones M, Marks LAB (eds) Disability, divers-ability and legal change. Martinus Nijhoff, London

Miller GP, Singer LS (2000–2001), Handicapped parking. Hofstra L Rev29:81

Pothier D, Devlin R (eds) (2006) Critical disability theory: Essays in philosophy, politics, policy and law. UBC Press, Vancouver and Toronto

Scotch R (2000) Models of disability and the Americans with Disabilities Act. Berkeley J Empl & Lab L 21

Sealy P, Whitehead PC (2004, pp. 249–257) Forty years of deinstitutionalization of psychiatric services in Canada: An empirical assessment. Can J Psychiatr 49

Stein MA (2003) The law and economics of disability accommodations. Duke L J 53:79

Stein MA (2007) Disability human rights. Cal L Rev 95

Silvers A (1998) Formal justice in disability, difference, discrimination: Perspectives on justice in bioethics and public policy

Scotch R, Schriner K (1997) Disability as human variation: Implications for policy. Ann Am Acad, AAPSS 549

Sodden A (ed) (2005) Advising the older clients. Lexis Nexis Butterworths, Markham, ON

ten Broek J, Matson FW (1966)The disabled and the law of welfare Cal L Rev 54:809

The Roeher Institute (1996, pp. 3–5) Disability, community and society: Exploring the links. The Roeher Institute, North York, ON

Chapter 8

Equity Theory: Responding to the Material Exploitation of the Vulnerable but Capable

M.I. Hall

8.1 Why "Law and Aging"? The Problem of Vulnerability

The very idea of "law and aging" as a discrete category of legal principle and theory is controversial: how and why are "older adults" or "seniors" or "elders" (the terminology is itself controversial and fraught with difficulties) a discrete and distinct group for whom "special" legal thought and treatment is justified? For some, a category of law and aging is inherently paternalistic, internalizing ageist presumptions through the suggestion that older persons *per se* are, like children, especially in need of the protection of the law. The consequences of that internalization are ultimately harmful. If certain older adults are genuinely in need of special legal protections because of physical or mental infirmities their need should be conceptualised in terms of their disability; older adults are not a distinct group but an arbitrarily delineated demographic category which contains within it any number of groups that are legitimately distinct for the purposes of legal theory (the disabled; women; persons of colour; Aboriginal persons; rich and poor; etc.) Indeed, the artificial category of "older adults" may be seen as obfuscating, submerging these more meaningful distinctions.

The essential question underlying any theory of law and aging is, therefore, this question of conceptual distinction. What special features and characteristics of "older adults" justify and even require a particular theoretical approach? Is it possible to formulate legitimate generalizations about a group identified as "older adults" while avoiding the harmful stereotypes of ageism? And what if anything is gained by that approach?

These important questions must inform the discussion but ultimately do not undermine the value and validity of law and aging altogether. I suggest that there are particular characteristics generally associated with the process of ageing in modern Western societies such as Canada (my own society), characteristics that may be considered within a general rubric of *vulnerability*, which require a focused theoretical approach. Vulnerability in the law and aging context has two major aspects: social vulnerability (vulnerability arising from non-participation in the workforce and from ageist social attitudes) and the personal vulnerability that frequently increases with the aging process. Age-related personal vulnerability includes both individual vulnerability and vulnerability in the context of intimate relationships.

I. Doron (ed.) *Theories on Law and Ageing: The Jurisprudence of Elder Law*,

Vulnerability associated with aging may arise in a number of ways and through the interaction of different sources of vulnerability (both personal and social). For many older adults, for example, the line between mental capacity and incapacity is not bright but gradual and spotty; this "grey zone" can last for years, a source of insecurity through diminished ability to care for one's self and exposure to exploitation by others. Physical changes associated with aging, while stopping short of "disability," may increase vulnerability by increasing physical or psychological dependence on others. Those who look or behave in ways identified as "old" will be more affected by social vulnerability than the younger-seeming individuals prominent in advertisements and other "pro- aging" visual messaging.

Vulnerability arises by reason of the interaction of these and other age-related factors in the context of an individual's total life situation, including relationships, education, experience, personality, and connection to society through paid work or otherwise. It follows that each of these factors, and the relationship between them, will have a distinct meaning and significance for each individual. Nevertheless, it is realistic to identify general trends while constructing a theoretical framework that will account for and give meaning to individual differences (as we recognize a theoretical category of "women and the law" without imagining a homogenous monolith of women as the essential precursor). As we age, our peer and family group (parents, siblings) diminishes and the social connections formerly supplied by employment or parenting generally fall away. The physical changes associated with aging will be frightening for many people, and the reduced sense of physical power will give rise to increased emotional vulnerability for many (although not necessarily all). There are also significant characteristics associated with the current generation of older adults that work to increase vulnerability. Older adults are more likely to have a lower level of education, for example. The rapid technological and social changes of the last two decades may be confusing, particularly for women within the current general of older adults who have spent their lives in traditional marriages without independent access to or control of economic resources.

Resistance to the idea of vulnerability as key to a conceptually coherent category of "law and aging" is strong, and rooted in the idea that vulnerability = weakness and resistance to the presumption that age = loss of capacity. The fear is that legal theory focusing on personal vulnerability increases *social* vulnerability, the more significant source of harm, to the extent that it reinforces ageist presumptions of weakness and incapacity. Legal protection for the truly incapable, of whatever age, exists; and beyond that, older adults should be treated in law and otherwise like any other adult persons.

Autonomy is a key value, and fundamental to our notion of legal personhood, but concern for autonomy should not subsume concern for the security and well-being of the vulnerable older adult. Personal and social vulnerability *do* make many older adults vulnerable to self-neglect, and to exploitation and abuse by other persons, even where the individual is determined to be mentally capable. As a society, do we set aside those concerns and conceptualise self-neglect and exploitation as choices, objectively harmful but nevertheless freely chosen by the individual and therefore outside the ambit of law's intervention? Does the protection of autonomy require, in effect, the editing out of the generalized social and personal vulnerability of older adults as legally irrelevant? After all, many people, perhaps all of us, are vulnerable to a certain extent across the age range, and our personal vulnerabilities are, generally speaking, not the concern of the law.

This is the legislative approach that has been taken by the Canadian province of British Columbia in its abuse and neglect legislation, the *Adult Guardianship Act (Part Three)*.[1] The Act defines "abuse" as the deliberate mistreatment of an adult that causes the adult

(a) Physical, mental or emotional harm, or
(b) Damage to or loss of assets,

and includes intimidation, humiliation, physical assault, sexual assault, overmedication, withholding needed medication, censoring mail, invasion or denial of privacy or denial of access to visitors.

Where abuse, as defined above, is established, the Act provides that the individual suffering from abuse is to be offered a response plan, which he or she is then free to accept or decline. If the plan is declined, an assessment will be made as to whether the adult is capable of declining that plan, in terms of his or her mental capacity.

This scheme is in keeping with and furthers the overarching principles as defined in the legislation, to recognize and protect individual autonomy. In service of this objective, however, the legislation codifies the policy decision to edit out vulnerability as a relevant consideration. The requirement of deliberation on part of an intentional wrongdoer works to exclude more ambiguous factual situations from the scrutiny of the law, unless the vulnerable individual can be re-categorised as incapable, a response that provides protection but is stigmatizing and may result in an unnecessary loss of independence. The presumed ability to "choose" situations of abuse or neglect ignores the relevance of relationship context, for example, to independent decision making; is the older woman who has spent her life in a traditional marriage, subordinating he own best interests to the interests of her husband and children, "freely choosing" to remain in an abusive relationship with a spouse whose abusive and controlling behaviour is escalating as he enters the "grey zone" (refusing to allow home help, for example, insisting that his wife carry out all domestic responsibilities herself)?

A significant theoretical issue for law and aging is whether and how vulnerability can be understood and situated within a theory of law and aging without internally replicating ageist presumptions that increase social vulnerability. If the task is impossible, perhaps the approach embodied in the British Columbia legislation, to edit out vulnerability short of incapacity as legally irrelevant, is necessary to protect autonomy.

My hypothesis is that the task is not impossible, however, but entirely within grasp. The key is to rethink our ideas of vulnerability in this context, drawing on the conceptual framework of equitable fraud: the venerable doctrines of undue influence, and unconscionability. Equity provides a coherent and sophisticated

[1] R.S.B.C. 1996, c. 6.

theoretical model for understanding vulnerability as both situational and relational, as opposed to the capacity/autonomy duality. Undue influence and unconscionability are related, but distinct. The inequity of the unconscionable transaction lies in one person's exploitation of the vulnerability of another. The inequity of undue influence lies in the effect of one person's "undue" influence of the ability of another to give free and independent consent. In neither scenario does the "weaker" party lack capacity; vulnerability arises through the power dynamics of the relationship together with factors of dependence and inequality.

8.2 Undue Influence

A key insight underlying the conceptual framework of undue influence is the *construction* of vulnerability, not as a constant and organic state of being but as situational, arising from the interplay between relationship context and personal characteristics in a particular situation. The legal significance of vulnerability, so defined, arises from its particular impact on consent, where the focus of inquiry is consent in the context of the relationship which is itself the source of vulnerability. The undue influence analysis of vulnerability as socially constructed and situational is a realistic and appropriate theoretical lens through which to view the nature of age related vulnerability and the harms that may arise from it, and may suggest an appropriate legal response to those harms.

"In everyday life people constantly seek to influence the decision of others";[2] some are more persuasive than others, some more gullible, and in the normal course of events the law will not intervene to even the playing field. We are all entitled to make mistakes, even those we later regret. Influence becomes "undue," for the purposes of equity, when it impairs or vitiates the ability of the one so influenced to give free or genuine consent. The doctrine has been described as "plaintiff sided" (see Birks and Chin 1995)[1] in that the focus of inquiry and locus of inequity is the plaintiff; equity is concerned with the plaintiff's experience of influence and its impact on his or her state of mind. The conduct and motives of the defendant are relevant only in-so-far as they produce this effect on the plaintiff. Undue influence may therefore be exerted both intentionally and unintentionally. Indeed, the good son who acts only in his mother's best interests, described by the Court as "honest, reliable, very reserved and anything but aggressive and demanding" may nevertheless exert undue influence over his mother through her "confidence and trust" in him, justified though that be.[3] Undue influence in this kind of circumstance arises through the internal dynamics of a certain kind of relationship, in which the "potential for domination" is "inherent in the relationship itself."[4]

[2] *Royal Bank of Scotland* v. *Etridge*, [2001] 4 All E.R. 449 (H.L.) at 1.

[3] See, *Zed v Zed* (1980), 28 N.B.R. (2d) 580 (QB).

[4] *Geffen* v *Goodman Estate*, [1991] 2 S.C.R. 353.

The intentional/unintentional distinction in undue influence is important, and conceptually key to the distinction between undue influence and unconscionability. The classic doctrinal distinction between different qualities of undue influence is not between intentional and unintentional, however, but between "actual" and "presumed." "Actual" undue influence is exerted intentionally to achieve the end sought, and includes manipulation, coercion or threat. "Presumed" undue influence refers to undue influence arising through the dynamics of the relationship in question; where the relationship between A and B is one in which the potential for domination is inherent in the relationship itself, a presumption of undue influence will arise where A confers a benefit on B. The presumption will be displaced if B can establish that, despite the relationship, A did in fact freely consent to the transaction, but this is for B to prove if B is to keep the benefit.

The presumption mechanism may seem harsh in the case of the blameless good son, for example, where the "potential for domination" has arisen through the laudable assistance provided by the son to his mother. The policy rationale for the rule was described by Sir Erich Sachs in the case of *Lloyd's Bank v Bundy*[5]: the court "interferes not on the ground that any wrongful act has been committed by the donee but on the ground of public policy and to prevent the relations which existed between the parties and the influence arising therefrom from being abused." Relationships giving rise to a presumption of undue influence are by their nature private and the "dominant figure" in that relationship is the only one in a position to prevent that influence from being "abused" ("abuse" in this context "mean[ing] no more than that once the existence of a special relationship has been established, then any possible use of the relevant influence is, irrespective of the intentions of the persons possessing it, regarded in relation to the transaction under consideration as an abuse- unless and until the duty of fiduciary care has been shown to be fulfilled or the transaction is shown to be truly for the benefit of the person influenced.)[6] It is for this reason both fair and just to require that dominant figure, where he or she has received a benefit in the context of that relationship, to take active steps to ensure that the transaction was indeed freely chosen.

Certain classes of relationships will always give rise to a presumption of undue influence; lawyer and client, for example, or parent and minor child. Outside of this limited class, other relationships may give rise to the presumption if, on close examination of a particular relationship, they contain broadly analogous qualities of trust and dependence. No more precise formula is appropriate; relationships in which one party develops a dominating influence over another are "infinitely various" and there is no substitute for a "meticulous examination of the facts".[7] As Sir Sachs explained in *Bundy*, "it is neither feasible not desirable to attempt to closely to define the relationship, or its characteristics, or the demarcation line showing the exact transition point where a relationship that does not entail the duty passes into one that does."[8]

[5] [1974] 3 All E.R. 757 (C.A.).

[6] [1974] 3 All E.R. 757 (C.A.).

[7] *National Westminster Bank v Morgan*, [1985] 1 AC 686 (HL) per Lord Scarman.

[8] *Bundy*, *Geffen* v *Goodman Estate*, [1991] 2 S.C.R. 353.

Lord Browne Wilkinson, in *Barclay's Bank v O'Brien*,[9] suggested a third class of relationship: those which would not always and automatically raise the presumption, but which are more likely to raise a presumption of undue influence than other kinds of relationships. The relationship at issue in *Barclay's Bank v O'Brien* was the relationship between husband and wife. The wife had provided a guarantee for a loan to her husband's business; the business had been unable to make its payments and the bank now sought to collect from the wife. The wife argued that the guarantee could not be enforced on the basis of undue influence. The undue influence in question had, in this case, been exercised by the husband (and not the bank itself, as successfully argued in *Lloyd's Bank v Bundy*).

Rejecting the argument that the wife/husband relationship should always give rise to a presumption of undue influence, Lord Browne Wilkinson acknowledged that, for several reasons, undue influence was more likely to be a relevant factor in this kind of relationship than in most others. The emotional and psychological ties inherent in the spousal relationship generally made it more likely that undue influence would arise in the particular relationship under consideration. The tendency for transactions within the context of intimate relationships to be entered into relatively informally without the same level of external or professional input as "stranger" transactions made it less likely that risks would have been stated accurately and that truly independent advice will have been sought. The question of whether a presumption of undue influence arose in these kinds of relationships should be approached by the Court, in the opinion of Lord Browne Wilkinson, with a "special tenderness," a sensitivity to the increased likelihood of the factors giving rise to the presumption within the relationship. Lord Browne Wilkinson suggested that courts apply this special tenderness approach not just to marital relationships but to other relationships of co-habitation, which would tend to share the same relevant characteristics, and to the relationship between "aged parent and adult child," citing to the case of *Avon v Bridger*.[10]

Can we say that the undue influence analysis applies in any particular way to older adults? A survey of the case law reveals numerous cases involving older adults in which undue influence is alleged.[11] Factors which will be relevant to an undue influence analysis generally will *often*, although not always, be factors that are associated with the social and physical consequences of the aging process in certain kinds of relationships. This understanding is no more ageist than the recognition that certain factors will be more likely to impact the lives of women than men, visible racial minorities than members of the racial majority, so long as that understanding is grounded always in the particular, what Lord Scarman described as "a meticulous examination of the facts" of each individual's situation.

---

[9] [1994] 1 A.C. 180 (H.L.).

[10] [1985] 2 All E.R. 281 (C.A.).

[11] See, *Kovach v Beardsley*, [1991] 6 WWR 113 at 120 (Alta. QB).

8.3 Unconscionability

Equity sets aside transactions that are deemed to be "unconscionable" on the basis of exploitation; in that the unconscionable transaction is the product of one person's exploitation of another, it would be inequitable to give it legal effect. If undue influence is "plaintiff-sided" unconscionability is a "defendant sided" doctrine; the inequity arises from the conduct of the defendant. There can be no unintentional unconscionability.

Unconscionability, like undue influence, is about vulnerability that is both constructed and situational, with relative weakness defined with reference to the power balance of the relationship in question. Where an inequality between the parties puts one (the "weaker") in the power of the other (the "stronger"), and where the bargain is substantially unfair to the benefit of the stronger, a presumption of unconscionability will be raised This presumption will be rebutted if the defendant can show that the bargain was fair, just and reasonable.

Inequality in this context refers not to objective inequalities, such as disabilities (which may be understood as objective inequalities) but an inequality that, in the context of the relationship, impairs the individual's ability to protect his or her own best interests. A blind person, for example, may well be capable of protecting his or her interests and if so, will not be "weaker" in the sense that he or she is in the power of the other player in the relationship.[12] Chronic drunkenness, in comparison, is more likely to have the relevant psychological effect, striking at the ability of an individual to advance or protect his interests, and the defendant who knowingly takes advantage of that inability will be behaving unconscionably for the purposes of the doctrine.[13]

Age, like other objective "weakness," does not in itself create an inequality for the purposes of the doctrine, nor should age itself be an independent factor going to the balance of power between the parties. Factors or characteristics generally associated with the ageing process will be relevant in this context, however: loneliness and social dependence; fearfulness about the future (who will provide for me if not the "stronger" party?); a lack of knowledge regarding current business practices and values; the lack of assertiveness and confidence that may accompany physical decline; the occasional confusion and lack of clarity (short of incapacity) of the "grey zone." What must be shown, where the relevant weakness exists, is that the stronger party took advantage of that weakness to the detriment of the weaker. In the case of *Matheson Estate v Stefankiw*,[14] for example, the elderly transferee Matheson (now deceased) had transferred his property at a price below its actual value. The transferors were long term tenants on the farmland in question. Mr. Matheson showed several signs of objective weakness: as he grew older, he

---

[12] *Sperling Estate v Heidt* (1998), 178 Sask. R. 192 (QB), aff'd. (2000), 199 Sask. R. 256 (CA).

[13] See *Black v Wilcox* (1976), 12 OR (2d) 759 (CA).

[14] (2001), 191 Sask. R. 241 (QB).

became increasingly reclusive and fearful of dealing with strangers. His health and hygiene had deteriorated to a striking degree. He had begun to experience occasional delusional episodes- not sufficient, apparently, to lead to a conclusion of incapacity (and a conclusion that the transfer was void on that basis) but sufficient to place Mr. Matheson in the "grey zone." The tenants had provided assistance to Mr. Matheson over the years, and he was comfortable with them.

The Court concluded that, in the circumstances, Mr. Matheson had obtained something of real value through his transaction with the defendants aside from the price paid for his property. Mr. Matheson's gratitude to his tenants, and ability to express his gratitude through the transfer, had worth to him, and the transfer to his tenants also spared him the "agony" of having to deal with strangers. The tenants had not initiated the transfer- it was Mr. Matheson's idea- nor was there any evidence that they had pursued the transfer or in any other way sought to exploit their neighbour and landlord. Aware of Mr. Matheson's increasing difficulties, the tenants responded with assistance and support, rather than exploitation. Under the circumstances, the transaction was not unconscionable.

The *Matheson* case is important as an illustration of how the unconscionability analysis of power in relationships can play out in cases involving characteristics of "weakness" that are associated with aging generally. This is not to say that every individual will age in exactly the same manner as Mr. Matheson. Each person's story will, in this respect, be different. But we can- and in accordance with reality must- make some generalizations about the kind of characteristics that are typically associated with aging, in the same way that Lord Browne Wilkinson is justified in his conclusion that the marital relationship is more likely to give rise to a presumption of undue influence, while rejecting as unrealistic the idea that it *always* will.

Matheson Estate v Stefankiw also dramatizes the distinction between a "plaintiff sided" undue influence analysis and a "defendant sided" analysis of unconscionable transactions, as well as the points of similarity between the two. Most obviously, unconscionability asks us to consider the element of intentional exploitation which, while it may provide evidence of "actual" undue influence, is not the focus of the undue influence inquiry. Within each analysis, however, those elements of personal vulnerability which may give rise to a presumption of undue influence in a particular relationship will also be relevant in the relationship analysis for unconscionability (to show that the relatively weaker party is in the power of the relatively stronger). The relevant elements of constructed vulnerability- arising from both relationship context and personal characteristics- will tend to be the same. It is the ultimate meaning of that vulnerability within each analysis that is separate and distinct- the impact on the plaintiff's ability to consent, on the one hand (undue influence), and the exploitation it enables, on the other (unconscionability). Unfairness (in terms of the inequality of the bargain- Blackacre for $1) is for this reason essential to a finding of unconsionability, but of evidentiary value only for the purposes of the equitable analysis of undue influence (Blackacre for $1 may be evidence of undue influence, and manifest unfairness is essential to the presumption, but is not conceptually essential to the influence/consent analysis).

Perhaps most importantly, for the purposes of law and aging theory, both undue influence and unconsionability provide a conceptual framework for understanding vulnerability which is specific to each individual, turning on the particular facts of the individual's social situation, internal characteristics, and relationship context. Each analysis is sophisticated and particular, avoiding ageist generalizations ("the aged are doddery of mind"[15] or all older adults lack mental capacity to make decisions) while recognizing that characteristics associated with aging do create vulnerability and, on that basis, will often be a source of unfairness and harm. The doctrines are sufficiently flexible to allow for individual differences alongside this recognition, as in *Matheson Estate v Stefankiw*. The famous case of *Re Brocklehurst*,[16] considering alleged undue influence, provides another example. The case concerned a (now deceased) Lord, who had gifted his estate to a younger man who owned a nearby garage. Lord Brocklehurst and the garage owner had become friends, with the garage owner providing assistance in numerous ways to his older friends. Lord Brocklehurst emerges from the narrative of the case report as an eccentric "character," autocratic and strong willed; the court concluded that, in the context of this particular relationship, and despite Lord Brocklehurst's failing health and objective weakness and degree of physical dependence, the Lord was in fact the psychologically dominant figure and the garage owner his dutiful factotum.[17]

Equity provides us with a way of thinking about vulnerability as legally relevant that is particularly applicable, and appropriate, as a way of thinking about "law and aging" or law and aging theory. The autonomy/incapacity duality poses mentally capable adults as, by definition, autonomous and self-interested decision makers (the corollary being that the adult who is not "autonomous" in this sense is not capable). The vulnerability narrative provided through Equity describes something quite different, a "third way" of conceptualizing vulnerability itself, and of vulnerability as a matter for public policy. Theory's role here is to provide this conceptual underpinning is a way that is coherent and fair, and that allows us to see the full range of age related vulnerability without editing it out of the public sphere to be theorized as regrettable but essentially private suffering and loss.

Legal theory requires a careful analysis of the relationship between aging and vulnerability, but it is neither necessary, not necessarily desirable, that legislation incorporating that theoretical perspective be itself age specific; rather the analysis provided through legal theory can inform legislation of more general application in terms of how if can, will and should apply to the particular needs of older adults.

---

[15] As the presumption of undue influence was incorrectly characterized by Madam Justice Southin in *Longmuir v Holland* (2000), 192 DLR (4th) 62 (BCCA).

[16] [1978] 1 All ER 767 (CA).

[17] See also *Calumsky* v. *Karaloff,* [1946] S.C.R. 110; *Kits Estate* v. *Peterson* (1994), 161 A.R. 299 (Q.B.); *Zabolotney Estate Committee* v. *Szyjak* (1980) 5 Man. R. (2d) 107 (Q.B.); *Scott* v. *Clancy*, [1998] W.W.R. 446 (Sask. Q.B.); and *Tracy* v. *Boles* [1996] B.C.J. No. 52 (S.C.).

8.4 Equity Theory and Legislative Reform: Responding to Material Exploitation

The role of legal theory is to understand and explain reality and, in particular suffering and unfairness, in a way that is legally intelligible; to explain unfairness as injustice. Legal theory cannot, generally, be directly transferred into legal rules, but provides the necessary conceptual underpinning for legislative reform and common law development, a way of understanding that is capable of translation into "the law." Once vulnerability is "seen" through Equity's lens, for example, it becomes impossible to edit out as legally irrelevant. The next step is to give that theoretical understanding expression within the law itself.

Regarding private transactions, of course, the doctrines of equity have long operated as legal rules. In the common law, equity's conceptualization has informed the analysis of consent in the context of tort, outside of the transactional context.[18] The objective is to unpack those equitable rules in the particular context of law and aging, and then begin to consider how the resulting theory may inform other areas of the law; in particular, legislative responses to the problem of the material exploitation of older adults.

Consider the (hypothetical) story of Teresa X, an older woman who allows her 20-something grandson to move in "for a few weeks" until he gets back on his feet. A few weeks turn into a few months, and the grandson shows no signs of moving on. Grandson has issues with substance abuse, and shows no real interest in finding employment. Rather than looking for work grandson hangs about the house all day, increasingly with friends that make his grandmother feel uncomfortable. Mrs. X is supporting her grandson financially in addition to letting him live in her apartment. To sum, grandson's long term presence is not welcome and is having a negative impact on his grandmother's life.

A year passes, and grandson is still there. His substance abuse problems have intensified; he often behaves strangely, speaking harshly to Mrs. X. Mrs. X loves her grandson, but she has become rather afraid of him (this is something Mrs. X confesses only to herself, when feeling particularly low). She definitely fears and dislikes his friends, who she thinks are dealing drugs. In fact, she fears that grandson is involved in selling drugs also. Mrs. X is still supporting grandson financially, and has increasingly had to draw down her savings. Mrs. X finds it difficult to do full housekeeping but grandson does nothing and, between him and his friends, the apartment has become a shambles.

Mrs. X feels a great deal of shame about what has happened to her living situation-shame that her grandson is the way he is, shame that her home has deteriorated as it has, shame that she has been "too weak" (in her estimation) to stop this from happening. She worries obsessively about her dwindling financial nest egg.

Mrs. X would like grandson to change but, if that doesn't happen, she would like him to leave. She cannot bring herself to ask him to leave, however; she doesn't

[18] *Norberg v Wynrib* [1992] 2 SCR 226.

know how he would react and she is worried about his friends (who have come to use her apartment as a kind of base). Mrs. X feels she has no one to turn to. Her friends (never plentiful to begin with) no longer feel comfortable visiting. Mrs. X can sense that they feel intimidated by the constant presence of grandson and his friends, although grandson has never explicitly tried to exclude visitors. Meanwhile, Mrs. X's apartment manager has spoken to her about complaints received in connection with noise and late night comings and goings from her apartment. Mrs. X is terrified that she will be evicted from the apartment where she has lived for the last 25 years.

Mrs. X's grandson is now asking her for a "loan," to be supplied by Mrs. X cashing in her remaining RRSPs. He talks about it constantly; how a loan would allow him to invest in a "business opportunity" (which Mrs. X suspects involves drugs), finally get back on his feet, and find a place of his own. Grandson swears the money will be repaid within a year. When Mrs. X demurs, grandson appears crestfallen and says that no one in the family has ever had any confidence in him or his plans. Mrs. X feels very sad and guilty when she hears this; she suspects it to be true, and the root of grandson's problems. Mrs. X feels she has no choice but to go along with grandson's suggestion; the current situation is increasingly unbearable.

No crime has been committed here. Mrs. X's grandson is not stealing from her, he is not beating her. Nor would this situation be captured or fall within the definition of abuse in adult protection legislation such as British Columbia's *Adult Guardianship Act*; there is no evidence of intentional abuse. Grandson was invited into Mrs. X's home, and she has never asked him to leave. He has not threatened or physically intimidated her, nor has he acted to isolate her by disallowing visits from friends or other concerned individuals. Mrs. X's grandson has not cashed her cheques, nor has he forged her signature or deceived her. Indeed, grandson's motivations and intent in this situation may be ambiguous even to himself. He may have convinced himself that he is providing his grandmother with welcome and needed company, and that his business opportunity really will benefit them both, despite moments of guilt and self-doubt (easily removed through alcohol or other means). The pseudo-criminal abuse paradigm does not fit easily; "In a relationship in which one person is likely to want to give and the other is likely to feel an entitlement to receive, how can the law identify improper transactions?"[19]

Some American states have codified undue influence doctrine, to a certain extent, in legislation. Legislation in the State of Maine,[20] for example, provides that a presumption of undue influence arises where an elderly dependent person has transferred real estate or undertaken a major transfer of personal property or money [representing 10% or more of the elderly dependent person's estate] for less than full consideration to a person falling within a set of particular kinds of relationships,

---

[19] See Dessin (2000, p. 203, 213). Answering her own question, Prof. Dessin concludes (after a review of the case law) that judges "ratify transactions they view as 'normal' and invalidate those they perceive as 'abnormal'."

[20] 33 ME REV STAT ANN (West) δ 1022(1)(2001).

including family and fiduciary relationships. The transfer may then be set aside unless the older adult was represented in the transfer by legal counsel. An elderly dependent person is defined as a person aged 60 or older who is "wholly or partially dependent upon one or more other persons for care and support." How helpful is this straightforward codification of the doctrine to Teresa X? It would seem to add little to the equitable doctrines themselves, and pose the same practical problems for Mrs. X. The rule is applicable only to transactions and will not help Mrs. X with her most significant current problem, the continuing disruptive presence of grandson in her home, by reason of which she faces eviction. Furthermore the rule (like the doctrine) contemplates initiation by Mrs. X which is highly unlikely given the intimate relationship and psychological considerations involved.

Equity theory, drawing on but not strictly equivalent to the doctrines of equitable fraud, allows us to "see" Mrs. X's vulnerability as a source of inequity, or unfairness, justifying a legal response directed to the nature and cause of that unfairness. The appropriate and workable legal response will not reproduce the doctrines, but incorporate their analysis. Equity theory works here to explain the vulnerability of Mrs. X realistically, and without implying inferiority or lack of mental capacity. Seen through Equity's lens we understand *why* Mrs. X's situation justifies intervention. Notably, Mrs. X's vulnerability in and of itself is not the focus but, rather the harmful consequences of her vulnerability in this situation. Where (as in the case of Mrs X) harmful consequences are associated with circumstances evincing "undue pressure" intervention should follow without the requirement of (intentional) "abuse"; Mrs. X's ability to freely choose whether to accept or to reject services and/or assistance offered should be evaluated with regard to the pressure/influence dynamics of the relationship in question and not solely on the question of her mental capacity.

8.5 Conclusion

Equity theory offers a "third way" to respond to the material exploitation of the vulnerable but capable. Simply re-categorizing individuals as lacking capacity is conceptually inadequate. Whatever theoretical approach to capacity is applied, including a functional or situation specific approach, the capacity model misses the crucial constructive inter-play between relationship context and personal characteristics. Balancing the protection of the vulnerable with respect for autonomy and personal freedom need not, and should not, take place within a debate about defining capacity as medical definition or social/rights-based model. "Capacity", like "abuse", is conceptually inadequate to describe the scope of harms justifying legal response in situations of vulnerability.

The world is regrettably best by human unhappiness, ill-advised relationships, betrayal and loss; the great majority of these remain private and outside of law's proper ambit. How and why is law's response justified to protect the interests of an older woman like Teresa X? Is her situation essentially regrettable, but private, and her protection the business of her family and friends? The answer to this question requires a realistic appraisal of those private sources of protection in the current social context.

The story of modernity is, in part, a story about the flight of the individual from community generally and, in particular, from communal oversight and censure. Personal freedom and rights to privacy, defining values of liberal democracy, are the fruits of that process. Like all other goods, however, they have been purchased at a price. Community oversight also has value; most significantly, communal oversight provides important protection for vulnerable persons, providing both security and protection.

Nostalgia is predictable, but communal oversight and involvement can be restored as generally applicable social norms only at intolerable expense to individual freedom and autonomy. Unless one chooses to submit to the communal (through membership in a religious group, for example, or continued allegiance to a community of origin outside of the liberal paradigm) communal intimacy cannot be sustained in a social context in which the primacy of individualism is both encoded in culture and enshrined in law. This idea of community as optional (to be chosen or opted out of in favour of mainstream liberal ideology) is fundamental to modern liberal democracies such as Canada, effecting a balance between plurality and multiculturalism on the one hand and individual freedom rights on the other. The interests of the vulnerable but capable and, crucially, the question of whether protective legislation is justified, must be understood within this wider social context.

References

Peter Birks, Chin Nyuk Yin (1993) On the nature of undue influence In: Beatson, Friedman (eds) Good faith and fault in contract law. : London, UK

Carolyn Dessin (2000) Financial abuse of the elderly. Idaho L Rev 36:203

Chapter 9
Law and Aging: Mental Health Theory Approach

W.C. Schmidt

9.1 Purpose and Scope

The purpose of this chapter is to offer a mental health theory approach to law and aging. Key issues regarding legal capacity, mental competence, guardianship, and substitute decision making for older persons are presented as a core around which elder law is established. A critical perspective describes how legal concepts of capacity are used to conceal legal bias and ageism against the older population and contribute to their social exclusion. From a mental health perspective, elder law can serve as an emancipating assertion in an aging society.

After introduction and background, this chapter will take a critical perspective of law and aging covering: social control and the therapeutic state, social deviance and the medicalization of deviance, and aging as social deviance. Then the chapter will review the therapeutic state and therapeutic jurisprudence, including therapeutic jurisprudence in law and aging, and criticism of therapeutic jurisprudence in mental health law. An alternative mental health theory approach to law and aging is presented next, beginning with the normative premises and implications of free will and responsibility, and moving through the legal mental health system, the case for abolition of involuntary civil commitment, and sanism as a mental health theory approach to law and aging. The chapter ends with guardianship as a sanist, ageist archetype through a therapeutic jurisprudence analysis, guardianship outcomes studies, and a due process criticism of guardianship.

9.2 Introduction and Background

One of the first Kuhnian[1] imprimaturs, or coming of age, for the jurisprudence of elder law was *Elderlaw: Cases and Materials* in 1992, "the first edition of the first casebook on the emerging field of aging law and policy."[2] This first elder

[1] See Kuhn 1970, pp. viii, ix and 43. An index to paradigm shifts in the structure of scientific revolutions includes revelation through study of textbooks and research publications.

[2] See Frolik and Barnes 1992, p. vii. This casebook is in its fourth edition published in 2007.

I. Doron (ed.) *Theories on Law and Ageing: The Jurisprudence of Elder Law*,

law casebook was a lunch table progeny of the July 1988 conference on guardianship, a key elder law issue, at Frank Lloyd Wright's Wingspread in Racine, Wisconsin.[3]

However charming a birth at Frank Lloyd Wright's architectural masterpiece, the origins of aging law and policy are readily traceable in history to aging policy in England and the *Decretum*, a codification of Church law by an Italian monk, Gartian, in 1140 (Williamson et al. 1985, p. 39). The *Decretum*, applicable in England through medieval monasteries that owned 15% of the land, was a systematic statement of poor relief policy that distinguished for the first time between the elderly and other categories of poor (Williamson et al. 1985, p. 38).

9.3 A Critical Perspective of Law and Aging

9.3.1 Social Control and the Therapeutic State

One critical view of law and aging, or elder law, sees a dramatic historical shift of public policy about the elderly from concerns about equity and social justice to efficiency and cost containment (Williamson et al. 1985, p. 38). While Social Security, Medicare, and Medicaid achieved a sea change in the incidence of poverty and quality of life for the elderly, the preoccupations with cost containment in the late twentieth century arguably changed the function and political economy of elder law and policy from social justice to social control.

> Probably the most important point made by theorists of social control is that the nature or face of control has changed over the past 100 years; not only are targets told they are responsible for their plight, they are also led to believe that their loss of autonomy and relegation to others' care is for their own good [references omitted]. Essentially, social control mechanisms have been altered in two major, yet overlapping ways. First, enlightened bureaucratic management, through the use of policy and law, has replaced informal controls of the primary group and overtly coercive powers of the state as the most pervasive sources of control [references omitted]….
>
> Second, the major agents of social control are no longer sweatshop employers or law enforcement agents. Rather the benign bureaucratic model overlaps with a medical model of control, and psychiatrists and other medical personnel play a major part in defining people's possibilities in life [reference omitted]…. [M]ental hygiene is not so much a science for preventing mental disorder as a science for the prevention of moral delinquency [reference omitted].[4]

---

[3] See Frolik and Barnes 1992, p. vii. See also ABA Commission on the Mentally Disabled & Commission on Legal Problems of the Elderly, *Guardianship: An Agenda for Reform – Recommendations of the National Guardianship Symposium and Policy of the American Bar Association* (ABA 1989) [Wingspread Conference Facility, Racine, Wisconsin].

[4] See Williamson et al. 1985, p. 32. Williamson, Shindul, and Evans account for the extent of social control experienced by older persons from historical, cultural, and social perspectives.

Kittrie identified a "new hybrid system of social controls" which he designated as the therapeutic state (Kittrie 1971, p. 40). The therapeutic state is distinct from the welfare state, which includes aid to the poor, public education, housing assistance, retirement benefits, medical care, and possibly guaranteed income (not to mention corporate welfare) (see Kittrie 1971, p. 10, 41; see generally Myrdal 1967). "The welfare state makes services available to voluntary consumers; the therapeutic state assumes that its clients are too incompetent to be voluntary or to realize the beneficence of the proffered assistance and therefore attempts to administer its services involuntarily" (Schmidt et al. 1981, p. 9; Kittrie 1971, p. 41).

9.3.2 Social Deviance and the Medicalization of Deviance

Associated with social control is social deviance, whether "reciprocal, interdependent," and an "inevitable" element of the social system, or the "result of social control" (Williamson et al. 1985, p. 28). Traditional social "deviants" have included: those subject to divestment from the criminal law (e.g., incompetents to stand trial, insanity acquittees, incompetents to serve a sentence); the "mentally retarded"; the mentally ill; juvenile and "defective" delinquents; psychopaths; drug addicts; alcoholics; and the eugenically sterilized (see Kittrie 1971, p. 40; Miller 1980; see generally Parsons 1951). Continued and more recent social deviance includes: mental illness; alcoholism; opiate addiction; delinquency, hyperactivity, and child abuse; homosexuality; the born criminal; AIDS, domestic violence, co-dependency, and learning disabilities; and eating disorders, compulsive gambling, transsexualism, menopause, premenstrual syndrome, infertility, suicide, impaired physicians, post-traumatic stress disorder, and obesity (Conrad and Schneider 1992), as well as andropause, baldness, and erectile dysfunction (Conrad 2007), and genetic mutation, malfunction, and enhancement (Conrad 2005).

Dangers of the therapeutic state lie in the conditioning of society to consider those with a label of deviance as " 'different,' rarely considering the possibility that deviance could easily be broadened to encompass many unsuspecting candidates" (Kittrie 1971, p. 361).

> The danger... exists that in the implementation of the rehabilitative ideal, the social-defense role will gain ascendancy, leaving the individual with little or no protection from the powers of the therapeutic state.
>
> ... More and more persons will find themselves subject to compulsory treatment for the well-being of society in general with little or no protection offered against error, oversights, or untoward infringements of privacy. At that time we would truly be near the Brave New World. (Kittrie 1971, p. 401)

With the Brave New World of the therapeutic state and the medicalization of deviance, constitutional safeguards and individual rights are circumvented or lost in the name of health (see Conrad and Schneider 1992, p. 257; Kittrie 1971; see also Symposium 1990).

9.3.3 *Aging as Social Deviance*

A critical view of law and aging, or elder law, sees growing old as a form of social deviance: "the elderly are punished by isolation and stigmatization for this 'deviant' act" (Williamson et al. 1985, p. 4; Estes and Binney 1989, p. 587; Lock 1984, p. 121). Disvalued persons like the handicapped and the old are told they are normal and encouraged to act like they are normal while social organization precludes normalcy and acceptance (Williamson et al. 1985, p. 29). For example, while Social Security achieved a sea change in the incidence of poverty and quality of life for the elderly, its passage also authoritatively established the elderly as a new category of deviants.[5]

> The functions of Social Security for stabilizing the social order and thus social control were many....: unemployment was reduced, old people were to support the economy through consumption, they were to serve as both positive and negative role models for others, and the political steam they had gathered in working for change was to be diffused. Less obvious consequences of this process were solidification of age norms and the creation of numerous administrators, caretakers, and experts ["an aging enterprise"] who were to make their livings, their profits, and their reputation off of the elderly (Williamson et al. 1985, p. 105, 109).

Instead of income support for all, Social Security, like Medicare later, achieved income support for some. "Social Security launched a process by which a primary deviation, old age, became a secondary deviation – one with definite role expectations, not the least of which was exit from the labor force" (Williamson et al. 1985, pp. 110–111). Loss of employment implies disability and moral deficiency (Williamson et al. 1985, p. 111). "[D]isability connotes an incapacity to perform role expectations and is very much in keeping with the medical model of social control" (Williamson et al. 1985, p. 111).

9.4 The Therapeutic State and Therapeutic Jurisprudence

One of the more recent manifestations of the therapeutic state is therapeutic jurisprudence. Therapeutic jurisprudence, "the role of the law as a therapeutic agent" (Wexler and Winick 1991, p. 8), studies "the use of the law to achieve therapeutic objectives" (Wexler 1990, p. 4). Therapeutic jurisprudence is an "antidote" to the judicial opinion based "doctrinal, constitutional, and rights-oriented" approach reached by mental health law (Wexler and Winick 1991, p. 3, 7). Traditional mental

[5] See Williamson et al. 1985, p. 105. Social Security is a social insurance program, a safety net of income support and maintenance, and of economic security. As the largest source of income for older persons, 88% of Americans over age 65 received benefits in 2005: 69% of these beneficiaries received over half their income from Social Security; 40% of these beneficiaries received over 90% of their income from Social Security; and 25% of beneficiaries receive their only income from Social Security. See, e.g., Frolik and Barnes 2007, pp. 151–161.

health law extended rights in constitutional criminal procedure to the mental health system, but allegedly grew "sterile" with increased social conservatism and changes in composition of the U.S. Supreme Court.[6] Therapeutic jurisprudence examines "the extent to which substantive rules, legal procedures, and the roles of lawyers and judges produce therapeutic or antitherapeutic consequences."[7] Therapeutic jurisprudence has produced a considerable literature in mental health law, as well as such other fields of law as criminal law, tort, and contract (see references cited by Perlin 1998, pp. 534–544 – cumulative supplement in 2007, pp. 141–154). Elder law has accumulated therapeutic jurisprudence insights (see, e.g., Kapp 2003; Marson et al. 2004, p. 71; Stolle 1996, p. 459).

9.4.1 Therapeutic Jurisprudence in Law and Aging

In "geriatric" therapeutic jurisprudence, for example, the scope of *The Law and Older Persons* concentrates on law protecting older persons' well-being and rights in "various precarious settings," but excludes therapeutic jurisprudence analysis of age discrimination, income maintenance like Social Security or disability, and in-kind benefit programs like Medicare and Medicaid.[8] These are conspicuous exclusions, or they are so obviously "therapeutic" to the overall quality of life and

---

[6] See Wexler and Winick 1991, pp. 4–5. Cf. Ennis 1971, p. 101 (standards and procedures for involuntary confinement for mental illness should be no less than for criminal defendants).

[7] See Wexler and Winick 1991, p. ix. A 9-cell diagram with therapeutic, neutral, and antitherapeutic on the vertical axis, and substantive law, legal procedure, and legal role on the horizontal axis encourages therapeutic jurisprudence conclusions about whether any particular law, procedure, or role is therapeutic, antitherapeutic, both, or neither. Wexler 1990, pp. 4–5. Cf. the 4-cell matrix for social science in law with law, and social science, on the horizontal axis and substance, and method, on the vertical axis. Social science in law jurisprudence is a tool in law for analysis of the resulting four subtopics of: substantive law ("the legal rules which make the involvement [of social science in law] relevant"); legal method ("the process of managing the involvement [of social science in law"]; social science findings ("the relevant research results"); and social science method ("the [social science] techniques of carrying out and analyzing that research"). Monahan and Walker 2006, p. v. From a social science in law perspective, conceptual challenges for therapeutic jurisprudence include: legal method issues about the legally appropriate methods of using social science; social science method issues about research design (e.g., causation, internal and external validity); issues about the admissibility of social science findings at trial and on appeal as "adjudicative facts" or "social authority"; social science findings as "legislative facts"; and, social science findings as context or "social framework" for predicting future facts, determining present facts, and determining past facts. Social science in law jurisprudence may provide a more objective, neutral, or even determinative perspective than therapeutic jurisprudence. In contrast to the "analogical reasoning" of traditional legal doctrinal analysis, the typical "creative/analytical process" for therapeutic jurisprudence includes an introduction, a description of the pertinent law, a section on pertinent psychology, an integrative section applying or"relating the psychology to the law," and a conclusion. Wexler and Winick 1991, pp. 13–14.

[8] See Kapp 2003, p. 4. The precarious settings assessed by Kapp include guardianship and elder mistreatment, nursing homes, home and community-base care, consumer choice in health and long term care, research involving older participants, and regulation of the dying process. To

health for older persons as to presumably render analytic results a foregone conclusion.[9] However, any sense from Kapp that elder law and policy is generally "antitherapeutic" seems belied by his exclusion of the substantial portion of elder law doctrine and casebooks that addresses age discrimination, income maintenance like Social Security or disability, and Medicare and Medicaid.[10]

While the founders of therapeutic jurisprudence deny (intentions) that therapeutic jurisprudence supports or calls for a return to the therapeutic state,[11] significant criticisms of therapeutic jurisprudence at least keep the concern alive.[12]

---

achieve "quality assurance and elder rights objectives", Kapp advocates "first, do no harm" and comparison of "directive… lawmaking" to alternatives strategies like "market mechanisms (i.e., consumer choice and economic empowerment), professional education, private accreditation (e.g., Joint Commission on Accreditation of Healthcare Organizations), privately sponsored quality assurance (QA) endeavors, and innovative payment systems that create positive incentives for desired kind of behavior." Kapp 2003, pp. 6–7. The effectiveness of market mechanisms is at least controversial for elder rights and quality assurance. Compare, e.g., Kapp 2000, p. 3 with Schmidt 2000, p. 53. Private accreditation attracts significant criticism. See, e.g., Freeman 2000, p. 543; Furrow 1998, p. 361; Jost 1983, p. 835; 1994, p. 15; Kinney 1994, p. 47; LeGros and Pinkall 2002, p. 189; Metzger 2003, p. 1367; Michael 1995, p. 171; Symposium on private accreditation in the regulatory state 1994, p. 1; DHHS 1999, p. 2 (JCAHO surveys are "unlikely to detect substandard patterns of care or individual practitioners with questionable skills"). See also Schmidt et al. 2007, p. 641 (83% of guardians with high school or General Equivalency Diploma education are likely to have more severe certification sanctions compared with 76% with undergraduate or higher education, and 48% with an Associate of Arts degree or technical degree).

[9] But cf., some therapeutic jurisprudence could assert such psychological considerations as the stigma and loss of autonomy associated with needing age discrimination protection, and the dependency, stigma, and loss of autonomy associated with receipt of social security, disability, medicare or medicaid. Kapp identifies a purpose of extending therapeutic jurisprudence inquiries to ask whether legal involvement and intervention in older lives is a good thing not only for "the intended beneficiaries," but also "society as a whole." Kapp 2003, p. viii. Some therapeutic jurisprudence could assert that age discrimination statutes, social security, disability, medicare, and medicaid are "antitherapeutic" for society, future generations, or those who do not receive such allocations. Therapeutic jurisprudence can assert opposing conclusions.

[10] See Dayton et al. 2007, pp. 125–274, 451–487 (discrimination, planning for retirement, and health care including long term care) of 727-page book; Frolik and Barnes 2007, pp. 85–385 (age discrimination in employment, income maintenance, health care, long term care) of 674-page book.

[11] See Wexler and Winick 1991, p. xi ("Let us, at the outset, emphasize that therapeutic jurisprudence does not embrace a vision of law or even of mental health law as serving exclusively or primarily therapeutic ends. We do not call for a return to the 'therapeutic state' or extol what Wexler once called 'therapeutic justice.'"); Wexler 1993, p. 759, 762. ("Therapeutic jurisprudence in no way supports paternalism, coercion, or a therapeutic state. It in no way suggests that therapeutic considerations should trump other considerations such as autonomy, integrity of the fact-finding process, community safety, and many more.")

[12] See Haycock 1994, p. 301, 315 ("But if therapeutic jurisprudence is construed as a shift from, or an alternative to, rights-based perspectives, then real risks exist."); Petrila 1993, p. 877, 890:

> First,… the authors assume "general agreement that, *other things being equal*, mental health law should be restructured to better accomplish therapeutic values." This assumption on its face is highly questionable. Criticisms of the "therapeutic state" are common in both popular and professional literature; the views of people who do not share the belief that law should be devoted to accomplishing therapeutic values should not simply be discounted.

9.4.2 Criticisms of Therapeutic Jurisprudence in Mental Health Law

Petrila criticizes therapeutic jurisprudence: (1) for assuming that therapeutic outcomes should have a dominant, or any, role in judicial decision making; (2) for representing that therapeutic jurisprudence is a new approach to mental health law issues; (3) for ignoring the "consumer/survivor movement" in assuming "general agreement that, *other things being equal*, mental health law should be restructured to better accomplish therapeutic values"; and (4) for significantly failing to question "*who decides*" whether there is a therapeutic outcome and largely ignoring the people subjected to therapeutic jurisprudence (Petrila 1993). "Therapeutic jurisprudence as it has been conceptualized to date is a conservative, arguably paternalistic, approach to mental disability law" (Petrila 1993, p. 881). Petrila found no federal or state court cases using the term "therapeutic jurisprudence" (Petrila 1993. p. 878 n. 6).

Wexler and Winick respond by: (1) asking whether Petrila would have therapeutic and antitherapeutic consequences ignored; (2) agreeing that therapeutic jurisprudence is not new; (3) agreeing on the need "to be constantly vigilant in seeking out a patient/consumer perspective"; and (4) concluding that "therapeutic jurisprudence seeks to bring about a restructuring of mental health law more responsive to the interests, and desires, of its consumers" (Wexler and Winick 1993, pp. 907 n. 4, 909 n. 9 and 913–914).

Slobogin offers an identification and examination of five "conundrums" challenging therapeutic jurisprudence: "the identity dilemma"; "the definitional dilemma"; "the dilemma of empirical indeterminacy"; "the rule of law dilemma"; and, "the balancing dilemma" (Slogobin 1995, p. 193). Like Petrila, Slobogin assesses the extent to which therapeutic jurisprudence is new, or distinguishable from other legal philosophies sharing a goal to improve human well-being. In considering such other "jurisprudences" as legal realism, law and economics, law and society, law and psychology, social science in law, critical legal studies, feminist jurisprudence and feminist legal theory, and critical race theory, Slobogin concludes that any uniqueness (the identity dilemma) is emphasis rather than content. The definitional dilemma remains unresolved between therapeutic jurisprudence as a promoter of autonomy (the option of "the old libertarians"), a promoter of social adjustment, or "an academic term for 'happiness jurisprudence' " (the two options of "the old paternalists") (Slogobin 1995, pp. 203–204). For example, a right to refuse treatment consistent with a therapeutic value of choice may be antitherapeutic

---

Cf. Slobogin 1995, pp. 193, 211–215 (under the " 'Internal' Balancing" dilemma, while Wexler has asserted that therapeutic jurisprudence 'in no way supports paternalism, coercion, or the therapeutic state,' the logic of therapeutic jurisprudence "may obscure any values encapsulated in the Constitution not connected with therapeutic results" and "In short,... could undermine the normative premises of the legal system"). But cf. Wexler 1995, p. 220, 230 ("microanalytic therapeutic jurisprudence has in no way sought to construct a Therapeutic State: Therapeutic jurisprudence has been playing at the fringe of the mental health care tapestry and has not been designing the larger pattern"); Wexler and Winick 1993, p. 907.

in the event of psychological regression. Therapeutic jurisprudence can assert, and not readily resolve, opposing conclusions. Therapeutic jurisprudence "may often merely be putting old wine in new bottles" (Slogobin 1995, p. 204).

The dilemma of empirical indeterminacy reflects the reliance of therapeutic jurisprudence on social science, especially the challenges of ethical and constitutional limitations on randomization in experiments involving people, and of "the inherent tradeoff between internal and external validity" (Slogobin 1995, p. 203). The definitional dilemma of therapeutic jurisprudence exacerbates the empirical indeterminacy dilemma: "the social science generated by [therapeutic jurisprudence] may be unusually uncertain. If so, [therapeutic jurisprudence] will be relatively more speculative, for a longer period of time. In the meantime, its proposals may be hard to take seriously."

Regarding the rule of law dilemma, "the typical [therapeutic jurisprudence] article seems to be based on thought-provoking but speculative theories that are likely to be only partially supported by any good research that is carried out" (Slogobin 1995, p. 209). While the resulting law should be applied only "to those for whom it is therapeutic but not to those for whom it is not,… [g]iven our current state of predictive knowledge…, this approach could result in a considerable number of false positives and false negatives" (Slogobin 1995, p. 210). Slobogin concludes, "Although this dilemma afflicts any attempt at legal reform, it may be particularly excruciating for [therapeutic jurisprudence], given its quantitative empiricism. The choice is likely to be more informed, but also more painful" (Slogobin 1995, p. 210).

Slobogin defines the balancing dilemma as "How much weight should be given to showing that a legal rule or practice is therapeutic in light of countervailing considerations?" (Slogobin 1995, p. 210) The need for internal balancing occurs with the divergence of therapeutic values and other individual interests. "The empiricism of [therapeutic jurisprudence] and its alluring call for therapeutic results may combine to create a tendency to support proposals that, although not necessarily 'bad' in an ultimate sense, are thoroughly paternalistic" (Slogobin 1995, p. 212). Therapeutic jurisprudence "may obscure any values encapsulated in the Constitution not connected with therapeutic results" (Slogobin 1995, p. 213) and "could undermine the normative premises of the legal system" (Slogobin 1995, p. 214).

The need for external balancing occurs in weighing individual interests against the interests of others. Slobogin's principal point on external balancing is that a law's therapeutic value must not blind proponents "toward its potentially negative impact on others" (Slogobin 1995, p. 216).

In conclusion, Slobogin nonetheless hopes that therapeutic jurisprudence gives direction to the "notably rudderless" law and society, law and psychology, social science in law movement (Slogobin 1995, p. 219). "Therapeutic jurisprudence, carefully pursued, will help produce a critical psychology that will force policymakers to pay more attention to the actual, rather than the assumed, impact of the law and those who implement it" (Slogobin 1995, p. 219). In social science in law (and administrative law) terms, this seems to call for prioritizing enhancement of the use of social science findings as legislative facts, that is, social science findings about population or social epidemiological facts.

9.5 An Alternative Mental Health Theory Approach to Law and Aging

What alternative mental health theory approach to law and aging can be presented as a core around which elder law is established?

9.5.1 Normative Premises and Implications of Free Will and Responsibility

Compared with the deterministic assumptions of the behavioral sciences, the normative premises of law are free will and responsibility (Melton et al. 2007, pp. 8–9), including the legal fiction of law treating people for policy reasons as if they are responsible, even if scientifically or medically they are not responsible. Patients seek physicians who have deterministic assumptions and clients seek attorneys to represent client choices. Does a client want therapy services from an attorney? Does a patient want legal services from a physician? If an attorney does not represent the choices of an allegedly incapacitated person in a guardianship hearing, then who will?

In law, there is a legal presumption of innocence. The law so values liberty that it would rather let ten criminals go free than convict one innocent citizen.[13] The law errs on the side of freedom, committing figuratively, if not literally, ten times as many false negative errors as false positive errors. Accused criminals are held legally responsible only after the opportunities of notice and criminal trial requiring proof of guilt beyond a reasonable doubt. The alternative of presuming guilt and having to prove innocence is anathema to Anglo-American and Western industrialized law.

In analogous fashion, there is a legal presumption of competence, or capacity:[14] "Every human being of adult years and sound mind has a right to determine what shall be done with his body…"[15] The law so values liberty that it would rather let ten mentally incompetent people make decisions than adjudicate one competent person as legally incapacitated. The law errs on the side of freedom to make decisions, committing figuratively, if not literally, ten times as many false negative errors as false positive errors. Persons alleged to be incapacitated are adjudicated legally incapacitated and a guardian appointed to make their decisions only after the opportunities of notice and a hearing requiring proof of legal incapacity by clear

---

[13] "For the law holds, that it is better that ten guilty persons escape, than that one innocent suffer" (Blackstone 1783, p. 358).

[14] See, e.g., *In re* Boyer, 636 P.2d 1085 (Utah 1981) (mentally retarded persons are presumed legally competent to manage their personal and financial affairs); Doron 1999, p. 95, 104 (presumption of competency in common law and Ontario); Kapp 2003, p. 12; Perlin 1998, p. 529.

[15] Schloendorff v. Society of New York Hospital, 211 N.Y. 125, 129–130 105 N.E. 92, 93 (Cardozo, J. 1914).

and convincing evidence.[16] The alternative of presuming incapacity for adult human beings and having to prove competence is abhorrent to Anglo-American and Western industrialized law.

In the health care system, the legal normative premise is that an adult may consent or refuse any and all treatment, even if the consequence is death (see, e.g., Furrow et al. 2000, pp. 826–829 common law basis). The mental health system is normatively different.

9.5.2 *The Legal Mental Health System*

The mental health system "consists of the various agencies delivering mental health care and those portions of the judicial system with jurisdiction over proceedings to compel persons to submit to such care" (Miller et al. 1976, p. 1). The mental health system includes at least those aspects of the criminal justice system from which people move to mental health programs, including informal and formal diversion before and after arrest, incompetence to stand trial, insanity acquittees, guilty but mentally ill, sentencing option, transfers between correctional programs and mental health programs, and execution competency (see Miller 1976, pp. 1–2; Reisner et al. 2004). The mental health system also includes guardianship (Reisner et al. 2004, pp. 898–922). The existence of legal authority to compel care in the mental health system distinguishes the mental health system from the health care system (except for the overlap of consensual health and mental health care, and of health and mental health care secured through guardianship under the state's parens patriae authority).[17]

9.5.3 *The Case for Abolition of Involuntary Civil Commitment*

There is a strong case for abolition of involuntary civil commitment of the mentally ill, the legal authority to commit for treatment (see e.g., Chamberlain 1979; Dershowitz 1968, p. 71; Ennis 1972; Miller 1976; Reisner et al. 2004, pp. 682–686; Szasz 1968; Morse 1982, p. 54). Arguments against involuntary commitment include: (1) there is little justification for legally distinguishing mentally disordered persons from "normal" persons; (2) involuntary commitment produces unacceptable

[16] See, e.g., *In re* Boyer 636 P.2d 1085. Cf. Addington v. Texas, 441 U.S. 418 (1979) (the standard of proof for indefinite involuntary civil commitment to a state mental hospital must be greater than preponderance of the evidence, i.e., clear and convincing evidence).

[17] The constitutional validity of compulsory vaccination law, and the existence of forced treatment, isolation, and quarantine in public health law under the state's police power responsibility for public health are acknowledged. See, e.g., Jacobson v. Massachusetts, 197 U.S. 11 (1905); Furrow et al. 2004, pp. 87–112.

numbers of improper commitments (false positives); (3) states do not provide adequate treatment and care to people who are involuntarily committed; and (4) hospitalization is not necessary for most involuntarily committed persons.[18] There are also criticisms of outpatient involuntary commitment laws (Perlin 1998, pp. 497–498; Schwartz and Costanzo 1987, p. 1329; Stefan 1987, p. 288). Benefits of abolishing involuntary commitment include: (1) the expansion of liberty; (2) reduction of role confusion and resource waste for mental health professionals acting as agents of social control and healers; (3) treatment enhancement; and (4) cost savings (see Morse 1982, pp. 93–103).

The existence and implementation of a health system for people with mental illness that includes legal proceedings to compel people to submit to confinement and treatment is challenging as a viable mental health theory. One paradigm that provides some explanation, if not solace, is "sanism."

9.5.4 Sanism as a Mental Health Theory Approach to Law and Aging

Sanism includes the same kind of "irrational, unconscious, bias-driven stereotypes and prejudices" against people with mental illness that are exhibited in racism, sexism, homophobia, religious and ethnic bigotry, (and ageism) (Perlin 1998, p. 527; 1992, p. 373; Perlin and Dorfman 1993, p. 47). For example, one state supreme court cited sanism as a factor in considering legal counsel's effectiveness in involuntary commitment hearings: "The use of stereotypical labels – which, as numerous commentators point out, helps create and reinforce an inferior second-class of citizens – is emblematic of the benign prejudice individuals with mental illnesses face, and which are, we conclude, repugnant to our state constitution."[19] Furthermore, "Sanist decisionmaking infects all branches of mental disability law, and distorts mental disability jurisprudence. Paradoxically, while sanist decisions are frequently justified as being therapeutically based, sanism customarily results in antitherapeutic outcomes" (Perlin 1998, p. 527).

The sanist legal mental health system compelling people to submit to confinement and treatment, and involuntary civil commitment of people with mental illness, may one day prove as anachronistic as leper colonies, tuberculosis sanitariums, lawful racial segregation, apartheid, internment camps, threatened quarantine of persons with AIDS, and the like. Law reform tools like class actions against public mental institutions, civil rights and public interest law litigation, the Bazelon Center

[18] See Morse 1982, pp. 59–87. See also Reisner et al. 2004, pp. 682–686. There is significant research concluding that persons with mental illness have abilities to make decisions that are similar to persons without mental illness, and that mental illness is not synonymous with impaired decisionmaking capacity. See, e.g., Appelbaum and Grisso 1995, p. 105; Grisso and Appelbaum (1995), p. 149; Grisso et al. 1995, p. 127.

[19] Matter of Mental Health of K.G.F., 2001 MT 140, 29 P.3d 485, 495 (2001).

for Mental Health Law, Social Security disability benefits, the Rehabilitation Act of 1973, the Protection and Advocacy for Mentally Ill Individuals Act, the Americans with Disabilities Act, the Individuals with Disabilities Education Act, *O'Connor v. Donaldson*, *City of Cleburne, Texas v. Cleburne Living Center*, and *Olmstead v. L.C.*, make prove ultimately salutary.

Viewed in this light, the significance of mental health law concepts like legal capacity, mental competence, guardianship, and substitute decision making as a core around which elder law is established becomes more evident. These mental health law concepts, when used as instruments of sanism, if not of the therapeutic state, are reinforcing, congruent with, and complementary to, their use as tools of ageism, if not of the therapeutic state.

9.6 Guardianship: A Sanist, Ageist Archetype?

"Parens patriae" means parent of the country and is the sovereign role of the state as guardian of legally disabled persons such as "infants, idiots, and lunatics" (Black's 1979, p. 1003). Parens patriae is the responsibility of the state to take care of citizens who are unable to take care of themselves. In English common law, the King had the royal prerogative in the *de Praerogativa Regis* statute to act as guardian. This function belongs to the states, not the federal government, in the United States.[20]

A legal guardian historically is "A person lawfully invested with the power, and charged with the duty, of taking care of the person and managing the property and rights of another person, who, for defect of age, understanding, or self-control, is considered incapable of administering his own affairs."[21] In 1981, for example, most state guardianship statutes (Schmidt et al. 1981, p. 9) based the determination of incapacity on medical condition, or on sanist and ageist labels such as mental disability or advanced age. In 2005, while most guardianship statutes rely on functional incapacity, four states (Connecticut, New Jersey, Tennessee, Vermont) limit public guardianship services to those over age 60, four other states (Arkansas, Maryland, New York, Texas) limit services to those requiring adult protective

[20] "[T]he powers not delegated to the United States by the Constitution… are reserved to the States respectively…." U. S. Const. amend. X.

[21] See Black's 1979, p. 635. "[A]nyone, especially an older person, who needs a guardian is popularly assumed to be mentally ill. The aged person with a few of the symptoms of chronic brain syndrome, such as forgetfulness, is more likely to be judged mentally ill and therefore to be declared incompetent." Regan and Springer 1977, p. 36, a working paper prepared for the Special Committee on Aging, U.S. Senate (1977) citing Lehman 1962, p. 312. See also Schmidt et al. 1981, p. 9; Cohen 1967, p. 96; Report of the Task Panel on Legal and Ethical Issues to President's Commission on Mental Health 1978, p. 1395. The many studies since 1987 of adults under guardianship indicate that they tend to be "older adults, female, and have relatively low income." (Reynolds 2002, p. 109. See also Frolik and Barnes 1992, p. 783 (disproportionate number of elderly people subject to guardianship).

services (abused, neglected, or exploited), and four states (California, Maine, Ohio, South Carolina) focus services on specific mental disabilities.[22]

9.6.1 A Therapeutic Jurisprudence Analysis

The national movement to reform state guardianship laws beginning in the 1980's focused on tightening statutory criteria and procedures (Kapp 2003, p. 15). Conducting a therapeutic jurisprudence analysis, Kapp asserts there is a "significant challenge to any useful assessment" of the guardianship system's success that is grounded in a "basic divergence" in 2001 about the goals of the guardianship system (Kapp 2003, p. 17). He asks, "should a fundamental goal of a therapeutic guardianship system be to foster the creation of *many* new guardianships or, conversely, to minimize the use of this legal instrument and create as *few* new guardianships as possible?" (Kapp 2003, pp. 17–18) He argues a divergence between "guardianship *qua* therapy" with a focus on "guardianship as a means of protecting vulnerable persons against neglect and exploitation, on the one hand," and, on the other hand, an "adversarial due process" model of "guardianship *qua* limiter of individual rights… something to be used only as a last resort"… and… "a dangerous excuse for turning state benevolence into overreaching paternalism."(Kapp 2003, pp. 17–18)[23] Under the "therapeutic" conceptualization, the purpose of guardianship is to "protect and promote the well-being of seriously disabled persons who cannot fend for themselves in a perilous world, and to do so at the least economic cost and psychological cost as possible" (Kapp 2003, p. 18). Kapp says that the "therapeutic" model

> proposes that the guardianship system serve a therapeutic role, both in terms of facilitating rather than impeding the benevolent provision of helpful services to the incapacitated person and doing so with a minimum of unnecessary hassle and expense. In this model, courts are therapeutic agents more than neutral referees of disputed facts and technical points of law, while attorneys for the parties serve more as part of the caregiving team for the ward/potential ward than as combatants in search for decisive legal victory. Diversion of individuals to alternative arrangements short of formal guardianship proceedings, including extralegal "bumbling through" on the basis of trust and goodwill among the parties, would be part of the therapeutic theme (Kapp 2003, p. 18).

Fostering the creation of many new guardianships, making courts therapeutic agents, making attorneys part of the "caregiving team," and diverting individuals from legal proceedings to extralegal "bumbling through" seem consistent with a therapeutic state approach, and not in a good way, even, or especially, if successful. Many new guardianships is explicit social control and inconsistent with the least

[22] See Teaster et al. 2008, p. 28. Cf. Doron 1999, p. 110 (Ontario shifted to a capacity concept of guardianship that disconnected guardianship from paternalistic values, "mental health" attitude, and ageist stereotype).

[23] Cf. Schmidt 1995.

restrictive alternative constitutional principle.[24] Courts as therapeutic agents who "contract out" some of their sovereign guardianship monitoring tasks (Kapp 2003, p. 20) are not exactly the constitutional due process values of notice and an opportunity to be heard before a neutral judicial factfinder,[25] let alone the right to trial by jury.[26] But it does start answering Petrila's "who decides" criticism[27] of therapeutic jurisprudence by eliminating blindfolded Lady Justice as the iconic option. Attorneys as part of the "caregiving team," perhaps as agents of the therapeutic court/team, seem to facilitate legal and social exclusion, voicelessness, and disempowerment for the alleged incapacitated (presumed incapacitated?) person.[28]

Kapp appropriately addresses "data paucity regarding functional operation and impact of the guardianship system," including "[b]etter evidence of actual effects on real older people."[29] He recommends that the states' "continual tinkering with

---

[24] See, e.g., *In re* Boyer 636 P.2d 1085, citing Shelton v. Tucker, 364 U.S. 479, 488 (1960) ("The breadth of legislative abridgement must be viewed in the light of less drastic means for achieving the same basic purpose.").

[25] Cf. Suzuki v. Quisenberry, 411 F. Supp. 1113, 1128 (D. Hawaii) (neutral judicial officer required in involuntary commitment hearing); Bentley v. State *ex rel.* Rogers, 398 So.2d 992, 995 (Fla. Dist. App. 1981) (unconstitutional for hearing officer to determine competency of patient to consent to treatment, instead of court).

[26] The right to a trial by jury in guardianship cases has grown from eleven states in 1981 to 35 states in 2005. Teaster et al. 2008, p. 31.

[27] See Petrila 1993.

[28] The right to legal counsel in guardianship cases has grown from twenty-two states in 1981 to over twenty-five states in 2005, generally without charge to indigent respondents. Teaster et al. 2008, p. 30. Regarding the role of legal counsel as zealous advocate in guardianship cases, see, e.g., Perlin 1998, pp. 280–283 – cumulative supplement in 2007, pp. 92–96; Frolik 1981, pp. 599 and 634–635; Schmidt 1993, p. 39.

[29] Kapp 2003, p. 21, 26. Kapp also recommends such "[a]lternative interventions" as "case or care management services, money management programs, and representative payees" and "mediation" to achieve delay or diversion of guardianship. Kapp 2003, pp. 24–25. However, available studies do not so far support the hypothesized outcomes. See Wilber 1991, p. 150 (elegant quasi-experimental design showing daily money management does not seem to divert from conservatorship); 1996, pp. 213–224; Schmidt 2002, pp. 1029–1032 (regarding diversion); Teaster et al. 1999, pp. 141–149; Wilber 1995, p. 39; Wilber and Reynolds 1995, p. 248. For published empirical research that begins the identification of factors placing older adults at risk for conservatorship, see Reynolds and Wilber 1997, p. 87. Regarding mediation in adult guardianship, see, e.g., Radford 2002, p. 611, 615, 620 (stating that "There is little empirical evidence as to the use or effectiveness of mediation in guardianship cases" and that mediation in guardianship may jeopardize the rights of outsiders like women and minorities); Schmidt 2002, pp. 1032–1034 (women's satisfaction with mediation in divorce child custody is less than with adversarial proceedings); Wood 2001, p. 785, 808 ("Mediation is premised on the notion that the disputing parties understand the problem at issue and the process for resolution."). Tightening statutory criteria and procedures in guardianship empowers respondents and facilitates negotiation, if not mediation, as in, for example, criminal justice plea bargaining and in divorce and child custody disputes. Under a *Mathews v. Eldridge* analysis of guardianship, the private interests of rights lost and likelihood of institutionalization, and the high risk of erroneous decisions with less procedure, arguably substantially outweigh the government's interest in not so benevolent paternalism. Mathews v. Eldridge, 424 U.S. 319 (1976). See, e.g., note 38; text accompanying notes 30–35. Regarding problems with the annual accounting form and detecting and preventing misuse of funds by representative payees, see Report of the National Research Council of the National Academies 2007.

guardianship laws and practices should be informed and influenced by [therapeutic jurisprudence] considerations" (Kapp 2003, p. 26). However, Kapp offers scant data on positive impact and actual effects of therapeutic jurisprudence considerations on the guardianship system and real older people.

9.6.2 Guardianship Outcomes Studies

The professional literature and research on guardianship as of 1981 contained "very little that is supportive of general guardianship practice or that argues forcefully for guardianship" (Schmidt 2002, p. 10). The general absence of professional literature and research supporting guardianship or arguing forcefully for guardianship continues (Schmidt 1990, pp. 61–80; Schmidt 2002; Teaster et al. 2008).

A leader in health care quality assessment theory, Avedis Donabedian, identifies three major approaches to quality assessment: structure, process, and outcome (Donabedian 1980, pp. 79–84). Outcome evaluations have substantial advantages over structure and process measures of quality (Furrow et al. 2004, p. 22). While there is a lack of systematic studies (Wilber 1997, p. 272) in guardianship and adult protective services, the few outcome studies are important.

Blenkner and associates performed one of the first such studies through Cleveland's Benjamin Rose Institute. In the kind of quasi-experimental research design appropriately advocated by Slobogin (one of the constructive critics of therapeutic jurisprudence),[30] the experimental group receiving enriched protective services, including guardianship, not only failed to have deterioration or death averted, it also had a higher rate of institutionalization and death than the control group.[31]

The issues of the use of adult protective services, including pursuit of guardianship, and nursing home placement were not "revisited in an epidemiologically rigorous fashion" until 30 years later.[32] Linking the New Haven Established Population for Epidemiologic Studies in the Elderly cohort with catchment area adult protective services records and data from the state long-term care registry, Lachs and associates explored "whether APS use for abuse and self-neglect is an independent

[30] See Slobogin 1995, p. 203.

[31] See Blenkner et al. 1971, p. 483; 1974. A later reanalysis by other researchers suggested that the findings on death came from initial group differences not controlled by the random sampling, but confirmed the strong effect of membership in the experimental group on the institutionalization tendency. Berger and Piliavin 1976, 205. Cf., e.g., Davis and Medina-Ariza 2001 ("new incidents of abuse were more frequent among households that both received home visits and were in housing projects that received public education"). See also Schmidt 1986, p. 101; Schmidt et al. 1988, p. 125; Teaster et al. 1999.

[32] See Lachs et al. 2002, p. 734. Cf. Williamson et al. 1985, p. 33 ("Nursing homes present an unusually graphic example of bureaucratic, medical, profit-oriented, and labeling control mechanisms in operation").

predictor of [nursing home placement] after adjusting for other factors known to predict institutionalization (e.g., medical illness, functional disability, and poor social support.)."[33] The Lachs study found that "the relative contribution of elder protective referral [including "pursuit of guardianship"] to [nursing home placement] is enormous ["4- to 5-fold risk conferred by elder mistreatment and neglect"] and far exceeds the variance explained by such variables as dementia, functional disability, and poor social networks" (see Lachs et al. 2002, pp. 737–738). While the Lachs clinicians "observed that often nursing home placement resulted in dramatic improvements in quality of life that was apparent to all observers – including [adult protective services] clients themselves," the authors note, "it is remarkable that controlled studies of differential outcomes of [adult protective services] have not been conducted. A review of the literature shows no systematic attempt to evaluate program outcomes or to examine unintended consequences of [adult protective services] intervention" (See Lachs et al. 2002, p. 738).

Such adult protective services and guardianship outcomes research is consistent with the earliest descriptive research and conclusions about guardianship.

> Under the present system of "Estate Management by Preemption" we divest the incompetent of control of his property upon the finding of the existence of serious mental illness whenever divestiture is in the interest of some third person or institution. The theory of incompetency is to protect the debilitated from their own financial foolishness or from the fraud of others who would prey upon their mental weaknesses. In practice, however, we seek to protect the interests of others. The state hospital commences incompetency proceedings to facilitate reimbursement for costs incurred in the care, treatment, and maintenance of its patients. Dependents institute proceedings to secure their needs. Co-owners of property find incompetency proceedings convenient ways to secure the sale of realty. Heirs institute actions to preserve their dwindling inheritances. Beneficiaries of trusts or estates seek incompetency as an expedient method of removing as trustee one who is managing the trust or estate in a manner adverse to their interests. All of these motives may be honest and without any intent to cheat the aged, but none of the proceedings are commenced to assist the debilitated.[34]

A legal services advocate concluded: "When examined in the larger context of social programming through which we purport to help the less advantaged, involuntary guardianship emerges as an official initiation rite for the entry of the poor and the inept into the managed society…." (Mitchell 1978, p. 466) In short, "Recognize guardianship for what it really is: the most intrusive, non-interest serving, impersonal legal device known and available to us and as such, one which minimizes personal autonomy and respect for the individual, has a high potential

---

[33] See Lachs et al. 2002, p. 735. See also Reynolds and Carson 1999, p. 301 ("wards with family guardians were more likely to be living in the community than those with professional guardians").

[34] See Alexander and Lewin 1972, p. 136. See also Alexander 1977, p. 32. [In short, then, present California law and the law in most of the United States applies conservatorship: for inappropriate reasons (petitioner's unstated motives); according to invalid standards (old age, designing persons); under the dubious pretense of medical expertise; and without seeing to the representation of the proposed ward."]

for doing harm and raises at best a questionable benefit/burden ratio. As such, it is a device to be studiously avoided."[35]

9.6.3 *Due Process Criticism of Guardianship*

A product of a legal mental health theory application of the outcome studies to guardianship and adult protective services in elder law is a constitutional due process criticism. The landmark *Wyatt v. Stickney* mental health law case observed, "To deprive any citizen of his or her liberty [through involuntary civil commitment] upon the altruistic theory that the confinement is for humane, therapeutic purposes and then fail to provide adequate treatment violates the very fundamentals of due process."[36] This due process theory was affirmed by the Fifth Circuit which relied upon its articulation of a *quid pro* theory in *Donaldson v. O'Connor*:

> that civilly committed mental patients have a constitutional right to such individual treatment as will help each of them to be cured or improve his mental condition... that where the justification for commitment was treatment, it offended the very fundamentals of due process if treatment were not in fact provided; and... that where the justification was the danger to self or to others, the treatment had to be provided as the *quid pro quo* society had to pay as the price of the extra safety it derived from the denial of individual liberty.[37]

Thus, to deprive any citizen of his or her liberty through involuntary guardianship or adult protective services upon the altruistic theory that the confinement (in a nursing home, for example), or loss of all rights in plenary guardianship,[38] is for humane, therapeutic (jurisprudence) purposes, and then end up worse than a control

---

[35] See Cohen 1978. Guardianship represents a failure to execute appropriate planning and advance directives. Cf., e.g., Patient Self-Determination Act of 1990, 42 U.S.C.A. sections 1395(a)(1)(Q), 1395cc, 1396a (requires hospitals, skilled nursing facilities, home health agencies, hospice programs, and health maintenance organizations receiving Medicare and Medicaid to provide each patient with information about rights to accept or refuse treatment, to formulate advance directives, to document whether an advance directive is signed, to assure related state law is followed, and provide for education of staff and public about advance directives); Kapp 2003, pp. 143–156.

[36] Wyatt v. Stickney, 325 F. Supp 781, 785 (M.D. Ala. 1971), 334 F. Supp. 1341 (M.D. Ala.), 344 F. Supp. 373 (M.D. Ala.), 344 F. Supp. 387 (M.D. Ala. 1972), *aff'd sub nom.* Wyatt v. Aderholt, 503 F. 2d 1305 (5th Cir. 1974).

[37] Id., 503 F. 2d at 1312, *relying on* Donaldson v. O'Connor, 493 F. 2d 507 (5th Cir. 1974), *vacated*, O'Connor v. Donaldson, 422 U.S. 563 (1975).

[38] In most states, a finding of legal incapacity restricts or takes away the right to: make contracts; sell, purchase, mortgage, or lease property; initiate or defend against suits; make a will, or revoke one; engage in certain professions; lend or borrow money; appoint agents; divorce, or marry; refuse medical treatment; keep and care for children; serve on a jury; be a witness to any legal document; drive a car; pay or collect debts; manage or run a business. Brown 1979, p. 286. "The loss of any one of these rights can have a disastrous result, but taken together, their effect is to reduce the status of an individual to that of a child, or a nonperson. The process can be characterized as legal infantalization." Schmidt 1995, p. 6. See also Frolik and Barnes 2007, p. 443.

group, or end up serving third party interests, violates the very fundamentals of due process. The staffing ratios[39] and individual treatment and habilitation plans (see Schmidt et al. 1988; Teaster et al. 1999, p. 131, 149) remedies from *Wyatt* are appropriately incorporated as remedies in guardianship standards and practices.

The U. S. Supreme Court ultimately and importantly decided in the antisanist *O'Connor v. Donaldson* decision that persons (with mental illness) have a right to liberty: "a State cannot constitutionally confine without more [than the enforced custodial care received by Kenneth Donaldson for almost fifteen years] a nondangerous individual who is capable of surviving safely in freedom by himself or with the help of willing and responsible family members or friends."[40] This led to *Youngberg v. Romeo*, where the Court identified constitutionally protected Fourteenth Amendment due process rights of involuntarily institutionalized persons with mental retardation not only to "adequate food, shelter, clothing and medical care," but also to due process liberty interests in "conditions of reasonable care and safety," "freedom from bodily restraint," and "such minimally adequate or reasonable training to ensure safety and freedom from undue restraint."[41] The *O'Connor v. Donaldson* right to liberty and the *Youngberg* constitutional rights and liberty interests presumably apply to the legal mechanisms of guardianship and involuntary adult protective services.[42]

---

[39] See, e.g., Schmidt et al. 1981, p. 174 and 193: "public guardianship is being endorsed, but only if it is done properly. By 'properly' we mean with adequate funding and staffing, including specified staff-to-ward ratios…: no office of public guardian shall be responsible for more than 500 wards. When an office of public guardian is responsible for 500 wards, another office of public guardian shall be established for additional wards. No office of public guardian shall assume responsibility for any wards beyond a ratio of thirty wards per professional staff member. Schmidt et al. 1997, p. 23: First, the ratio of wards to public guardian should be revised from the previous suggestion of 30:1, particularly in light of the Virginia programs' inabilities to service short-term needs… The tendency of programs to ambitiously help clients in need is offset by services stretched too thin to adequately meets the needs of incapacitated persons. The ward to guardian ratio should be 20:1. Teaster et al. 2008, p. 94 ["Statutes in seven states (i.e., Florida, New Jersey, New Mexico, Tennessee, Vermont, Virginia, and Washington) provide for a ratio of staff to incapacitated persons served. These laws either require a specific statutory ratio or require that administrative procedures or contracts set out a ratio."] A class action in 1999 against a County Public Administrator providing public guardianship services alleged that the "guardian fails to engage sufficient numbers of professional personnel to be able to adequately assess and periodically reassess the needs of each of its individualized wards, to adequately formulate and periodically revise an individualized case plan for each of its wards, to insure the implementation of such case plans and to insure minimal professional interactions with each ward on an ongoing basis." Schmidt 2004, citing Tenberg v. Washoe County Public Administrator and Washoe County, No. CV99–01770 (Family Court, Second Judicial District Court, Nevada, filed March 15, 1999). The *Tenberg* case was settled. National Guardiaknship Association 2007 ("The guardian shall develop and monitor a written guardianship plan setting… short-term and long-term goals for meeting the ward's needs that are addressed in the guardianship order.…").

[40] 422 U.S. at 576.

[41] Youngberg v. Romeo, 457 U.S. 307 (1982).

[42] Cf. Americana Healthcare Corp. v. Schweiker, 688 F.2d 1072, 1086–1087 (7th Cir.), *cert. denied*, 459 U.S. 1202 (1983)(Youngberg standards apply to nursing home residents).

More recently, the U. S. Supreme Court found in *Olmstead v. L.C.* that the Americans with Disabilities Act ("the most important civil rights act ever passed by Congress to deal with problems of discrimination against person with disabilities") (Perlin 1998, p. 145) entitled state mental hospital residents to community-based integrated treatment instead of treatment in an unnecessarily segregated hospital, and that "Unjustified isolation is... properly regarded as discrimination based on disability."[43] *Olmstead* creates special implementation challenges and opportunities for older persons (Cohen 2001, p. 233).

Guardianship, then, is an archetype for therapeutic intentions in the therapeutic state potentially rationalized by therapeutic jurisprudence analysis. With guardianship system outcomes as yet unrealized, the need for the adversarial, due process model remains. Even, or especially, if guardianship system outcomes are realized, the need for the adversarial, due process model remains grounded in the normative premises and implications of free will and responsibility.

9.7 Conclusion

This chapter has offered a mental health theory approach to law and aging with a critical perspective of law and aging as a potential mechanism of socially controlling the social deviance of aging through the therapeutic state and therapeutic jurisprudence. An alternative mental health theory approach to law and aging is normatively premised on free will and responsibility and a legal mental health system that abolishes involuntary civil commitment and has sanism as a principal challenge. Guardianship was reviewed as a sanist, ageist archetype best perceived not through a therapeutic jurisprudence lens, but rather through a combination of guardianship outcomes studies and a due process criticism.

From a mental health perspective, elder law can serve as a social control mechanism of a therapeutic state and a therapeutic jurisprudence, or as an emancipating, empowering, integrating assertion in an aging society.

References

Alexander G (1977) Who benefits from conservatorship? Trial 13(5):30
Alexander G, Lewin T (1972) The aged and the need for surrogate management
Appelbaum P, Grisso T (1995) The MacArthur treatment competence study: I. Mental illness and competence to consent to treatment. L Hum Behav 19:105
Berger R, Piliavin I (1976) The effect of casework: a research note. Soc Work 21:205
Black's Law Dictionary (1979), 5th edn

---

[43] Olmstead v. L. C. *ex rel.* Zimring, 527 U.S. 581, 119 S. Ct. 2176, 2185 (1999).

Blackstone W (1783) Commentaries on the laws of England, 9th edn, book 4, chapter 27 (reprinted 1978)
Blenkner M, Bloom M, Nielson M (1971) A research and demonstration project of protective services. Soc Casework 52:483
Blenkner M, Bloom M, Nielson M, Weber R (1974) Final report: protective services for older people: findings from the Benjamin Rose Institute Study. Benjamin Rose Institute, Cleveland
Brown R (1979) The rights of older persons
Chamberlain J (1979) On our own: patient controlled alternatives to the mental health system
Cohen E (1967) Old age and the law. Women's L J 53:96
Cohen E (1978) Protective services and public guardianship: a dissenting view. 31st annual meeting of the gerontological society, Dallas, TX, 20 November 1978
Cohen P (2001) Being reasonable: defining and implementing a right to community-based care for older adults with mental disabilities under the Americans with disabilities act. Int J L Psychiatry 24:233
Conrad P (2005) The shifting engines of medicalization. J Health Soc Behav 46:3
Conrad P (2007) The medicalization of society: on the transformation of human conditions into treatable disorders
Conrad P, Schneider J (1992) Deviance and medicalization: from badness to sickness (expanded edition)
Davis R, Medina-Ariza J (2001) Results from an elder abuse prevention experiment in New York City. National Institute of Justice
Dayton AK, Wood M, Belian J (2007) Elder law: readings, cases and materials, 3rd edn
Dershowitz A (1967) Psychiatry in the legal process: a knife that cuts both ways. In: The path of the law from 1967: proceedings and papers at the Harvard Law School Convocation held on the 150th anniversary of its founding. Harvard University Press, A. E. Sutherland
DHHS (1999) Office of the Inspector General, external review of hospital quality: a call for greater accountability
Donabedian A (1980) The definition of quality and approaches to its assessment, vol. 1. Explorations in quality assessment and monitoring
Doron I (1999) From lunacy to incapacity and beyond – guardianship of the elderly and the Ontario experience in defining "legal incompetence". Health L Canada 19(4):95
Ennis B (1971) Civil liberties and mental illness. Crim L Bull 7:101
Ennis B (1972) Prisoners of psychiatry: mental patients and the law
Estes C, Binney E (1989) The biomedicalization of aging: dangers and dilemmas. Gerontologist 29:587
Freeman J (2000) The private role in public governance. N Y U L Rev 75:543
Frolik L (1981) Plenary guardianship: an analysis, a critique, and a proposal for reform. Ariz L Rev 23:599
Frolik L, Barnes A (1992) Elderlaw: cases and materials
Frolik L, Barnes A (2007) Elderlaw, cases and materials, 4th edn
Furrow B (1998) Regulating the managed care revolution: private accreditation and a new system ethos. Vill L Rev 43:361
Furrow B, Greaney T, Johnson S, Jost T, Schwartz R (2000) Health law, 2nd edn
Furrow B, Greaney T, Johnson S, Jost T, Schwartz R (2004) Health law: cases, materials and problems, 5th edn
Grisso T, Appelbaum P (1995) The MacArthur treatment competence study: III. Abilities of patients to consent to psychiatric and medical treatment. L Hum Behav 19:149
Grisso G, Appelbaum P, Mulvey E, Fletcher K (1995) The MacArthur treatment competence study: II. Measures of abilities related to competence to consent to treatment. L Hum Behav 19:127
Haycock J (1994) Speaking truth to power: rights, therapeutic jurisprudence, and Massachusetts Mental Health Law. New Eng J Crim Civ Confinement 20:301
Jost T (1983) The joint commission on accreditation of healthcare organizations: private regulation of healthcare organizations and the public interest. B C L Rev 24:835

Jost T (1994) Medicare and the joint commission on accreditation of healthcare organizations: a healthy relationship? Law Contemp Probs 57:15

Kapp M (2000) Health care in the marketplace: implications for decisionally impaired consumers and their surrogates and advocates. Ethics L Aging Rev 6:3

Kapp M (2003) The law and older persons: is geriatric jurisprudence therapeutic?

Kinney E (1994) Private accreditation as a substitute for direct government regulation in public health insurance programs: when is it appropriate? Law Contemp Probs 57:47

Kittrie N (1971) The right to be different: deviance and enforced therapy

Kuhn T (1970) The structure of scientific revolutions, 2nd edn (enlarged)

Lachs M, Williams C, O'Brien S, Pillemer K (2002) Adult protective service use and nursing home placement. Gerontologist 42(6):734

LeGros N, Pinkall J (2002) The new JCAHO patient safety standards and the disclosure of unanticipated outcomes. J Health L 35:189

Lehman V (1962) In: Kaplan J, Aldridge G (eds) Guardianship in social welfare of the aging

Lock M (1984) Licorice in leviathan: the medicalization of the care of the Japanese elderly. Culture Med Psychiatry 8:121

Marson D, Huthwaite J, Hebert K (2004) Testamentary capacity and undue influence in the elderly: a jurisprudent therapy perspective. L Psychol Rev 28:71

Melton G, Petrila J, Poythress N, Slobogin C (2007) Psychological evaluations for the courts: a handbook for mental health professionals and lawyers, 3rd edn

Metzger G (2003) Privatization as delegation. Col L Rev 103:1367

Michael D (1995) Federal agency use of audited self-regulation as a regulatory technique. Admin L Rev 47:171

Miller K (1976) Managing madness: the case against civil commitment

Miller K (1980) The criminal justice and mental health systems: conflict and collusion

Miller F, Dawson R, Dix G, Parnas R (1976) The mental health process, 2nd edn

Mitchell A (1978) Involuntary guardianship for incompetents: a strategy for legal services advocates. Clearinghouse Rev 12:451

Monahan J, Walker L (2006) Social science in law: cases and materials, 6th edn

Morse S (1982) A preference for liberty: the case against involuntary commitment of the mentally disordered. Calif L Rev 70:54

Myrdal G (1967) Beyond the welfare state

National Guardianship Association (2007) Standards of practice, 3rd edn

Parsons T (1951) The social system

Perlin M (1992) On "Sanism". SMU L Rev 46:373

Perlin M (1998) Mental disability law: civil and criminal, 2nd ed

Perlin M, Dorfman D (1993) Sanism, social science, and the development of mental disability law jurisprudence. Behav Sci L 11:47

Petrila J (1993) Paternalism and the unrealized promise of essays in therapeutic jurisprudence: a review of essays in therapeutic jurisprudence. N Y L Sch J Hum Rts 10:877

Radford M (2002) Is the use of mediation appropriate in adult guardianship cases? Stetson L Rev 31:611

Regan J, Springer (1977) Protective services for the elderly

Reisner R, Slobogin C, Rai A (2004) Law and the mental health system: civil and criminal aspects, 4th edn

Report of the National Research Council of the National Academies (2007) Improving the social security representative payee program: serving beneficiaries and minimizing misuse. National Academy Press, Washington

Report of the Task Panel on Legal and Ethical Issues to President's Commission on Mental Health (1978) Report to the President, Vol. IV, appendix

Reynolds S (2002) Guardianship primavera: a first look at factors associated with having a legal guardian using a nationally representative sample of community-dwelling adults. Aging Mental Health 6(2):109

Reynolds S, Carson LD (1999) Dependent on the kindness of strangers: professional guardians for older adults who lack decisional capacity. Aging Mental Health 3(4):301

Reynolds S, Wilber K (1997) Protecting persons with severe cognitive and mental disorders: an analysis of public conservatorship in Los Angeles county, California. Aging Mental Health 1(1):87

Schmidt W (1986) Adult protective services and the therapeutic state. L Psychol Rev 10:101

Schmidt W (1990) Quantitative information about the quality of the guardianship system: toward a next generation of guardianship research. Prob L J 10:61–80

Schmidt (1993) Accountability of lawyers in serving vulnerable, elderly clients. J Elder Abuse Neglect 5(3):39

Schmidt W (1995) Guardianship: court of last resort for the elderly and disabled

Schmidt W (2000) Health care financing and delivery for the elderly: a planned, regulated system in counterpoint to a competitive marketplace approach. Ethics L Aging Rev 6:53

Schmidt W (2002) The wingspan of wingspread: what is known and not known about the state of the guardianship and public guardianship system thirteen years after the wingspread national guardianship symposium. Stetson L Rev 31(3):1027, 1029–1032

Schmidt W (2004) Legal framework for evaluating public guardianship in virginia. 57th annual scientific meeting of the Gerontological Society of America, Washington, DC, 22 November 2004

Schmidt W, Miller K, Bell W, New E (1981) Public guardianship and the elderly

Schmidt W, Miller K, Peters R, Loewenstein D (1988) A descriptive analysis of professional and volunteer programs for the delivery of guardianship services. Prob L J 8:125

Schmidt W, Teaster P, Abramson H, Almeida R (1997) Second year evaluation of the Virginia guardian of last resort and guardianship alternatives demonstration project

Schmidt W, Akinci F, Wagner S (2007) The relationship between guardian certification requirements and guardian sanctioning: a research issue in elder law and policy. Behav Sci L 25(5):641

Schwartz S, Costanzo C (1987) Compelling treatment in the community: distorted doctrines and violated values. Loyola L Rev 20:1329

Slobogin C (1995) Therapeutic jurisprudence: five dilemmas to ponder. Psychol Pub Pol'y L 1:193

Stefan S (1987) Preventive commitment: the concept and its pitfalls. Mental Physical Disability L Rep 11:288

Stolle D (1996) Professional responsibility in elder law: a synthesis of preventive law and therapeutic jurisprudence. Behav Sci L 14:459

Symposium (1990) Challenging the therapeutic state: critical perspectives on psychiatry and the mental health system. J Mind Behav 11:247

Symposium on private accreditation in the regulatory state (1994) Law Contemp Probs 57:1

Szasz T (1968) Law, liberty and psychiatry

Teaster P, Schmidt W, Abramson H, Almeida R (1999) Staff service and volunteer staff service models for public guardianship and "alternatives" services: who is served and with what outcomes? J Ethics L Aging 5(2):131

Teaster P, Wood E, Schmidt W, Lawrence S (2008) Public guardianship after 25 years: in the best interest of incapacitated people? University of Kentucky Graduate Center for Gerontology and ABA Commission on Law and Aging, Lexington

Wexler D (1990) Therapeutic jurisprudence: the law as a therapeutic agent

Wexler D (1993) New directions in therapeutic jurisprudence: breaking the bounds of traditional mental health law scholarship. N Y L Sch J Hum Rts 10:759

Wexler D (1995) Reflections on the scope of therapeutic jurisprudence. Psychol Pub Pol'y L 1:220

Wexler D, Winick B (1991) Essays in therapeutic jurisprudence

Wexler D, Winick B (1993) Patients, professionals, and the path of therapeutic jurisprudence, a response to Petrila. N Y L Sch J Hum Rts 10:907

Wilber K (1991) Alternatives to conservatorship: the role of daily money management services. Gerontologist 31:150

Wilber K (1995) The search for effective alternatives to conservatorship: lessons from a daily money management diversion study. J Aging Soc Pol'y 7(1):39
Wilber K (1996) Commentary: alternatives to guardianship revisited: what's risk got to do with it? In: Smyer M, Schaie KW, Kapp M (eds) Older adults' decision-making and the law. Springer
Wilber K (1997) Choice, competence, and competency: the coming of age of protective services research. Gerontologist 37:272
Wilber K, Reynolds S (1995) Rethinking alternatives to guardianship. Gerontologist 35:248
Williamson J, Shindul J, Evans L (1985) Aging and public policy: social control or social justice?
Wood E (2001) Dispute resolution and dementia: seeking solutions. Ga L Rev 35:785

Chapter 10
The Future of Elder Law

R.C. Morgan*

Elder Law is unique in that it is one of the areas of law that is defined by the client (the age of the client) rather than the subject in defining the practice area.[1] Similarly, "[d]isability law is... defined primarily by the identification of a group within society as opposed to an area of law defined primarily by the cohesion of the substantive rules within the area–like contract law or tort law" (see Surtees, Chap. 7 at p. 94). The National Academy of Elder Law Attorneys (NAELA)[2] defines the practice of elder law as "[r]ather than being defined by technical legal distinctions, elder law is defined by the client to be served. In other words, the lawyer who practices elder law may handle a range of issues but has a specific type of clients – seniors."[3]

As noted by Professor Hall in Chap. 8, there is still some uncertainty about the recognition of elder law as more of an age-based practice (see Hall Chap. 8). Although more widely recognized as a practice area, at least in the United States, it is still a challenge to decide how to define it and variations on what to call it: later life planning, life care planning, elder law, law and aging or others. As Professor Dayton describes it:

One can describe the field as encompassing a broad range of matters ranging from health care access to age- and disability based discrimination, elder abuse, consumer fraud aimed at the elderly, wills, trusts, and estates, surrogacy, guardianships and conservatorships, and mental health law. It is probably more useful to describe elder law as the particular manner in which any aspect of law touches the lives of older persons. Some laws and policies are developed particularly out of concern for the elderly – public pension systems are implemented to assure or at least contribute to the income security of the aged. Other legal institutions are

* This chapter is written from the perspective of U.S. elder law, since the author is from the U.S. Some of the thoughts expressed in this chapter may have application to other countries.

[1] Another area of practice that has laws based on age involves children, whether it be delinquency, dependency, adoption, or guardianships or conservatorships of minors.

[2] See, http://www.naela.com/About_WhoWeAre.aspx?Internal = true (visited February 5, 2008).

[3] *See,* http://www.naela.com/Public_WhatIsElderLaw.aspx?Internal = true (visited February 5, 2008).

age-neutral in their reach – universal health care, for example – but are of special significance to older persons due the greater need of the elderly for certain kinds of health care. In short, some areas of law have a greater significance to older persons and affect the elderly population more profoundly than other demographic groups (Dayton Chap. 4 at 50–51).

The view of defining the practice by the clients served, rather than the subject matter, can have significant results. For example, an attorney in the United States who practices Elder Law could be expected to have an expertise in a significant number of areas of law[4] which might impact elders. An elder might, for example, commit crimes. Does that mean an Elder Law attorney would be the best choice to represent the elder defendant in a criminal proceeding rather than a criminal defense attorney? Perhaps not, but, would a criminal defense attorney have comparable knowledge of some of the social and health issues that may have contributed to the client's commission of a crime?

One would expect Elder Law attorneys in the United States to be well-versed in Medicare and Social Security laws and regulations. Many Elder Law attorneys have at least a passing understanding of those two subject matter areas but might be far less able to represent a client in a hearing before the appropriate government agency. Why? To oversimplify,[5] the laws and regulations are complex and if an attorney does not handle these cases frequently, it may be impractical to attain and maintain the necessary expertise. So, perhaps, as a result of defining a practice area by age, an interesting situation occurs: an Elder Law attorney may not be able to handle all legal problem an elder faces.

Elder Law attorneys have come to appreciate the importance of defining the strengths and focus of their practices in order to well serve their clients. This means that Elder Law has grown from a specialty to a general practice area within which attorneys may specialize. It is Elder Law having come of age in the United States (see generally, Needham, 2008). As Alan Bogutz writes in Chap. 1: "[l]onger life for a larger proportion of the population has been only one of the issues that led to the development of an area of law dedicated to the legal issues of aging" (Bogutz Chap. 1 at 2).

---

[4] The National Elder Law Foundation, or NELF, the certifying organization in elder law in the United States, defines elder law (and practice areas) as:

[t]he legal practice of counseling and representing older persons and their representatives about the legal aspects of health and long term care planning, public benefits, surrogate decision-making, older persons' legal capacity, the conservation, disposition and administration of older persons' estates and the implementation of their decisions concerning such matters, giving due consideration to the applicable tax consequences of the action, or the need for more sophisticated tax expertise.

[I]n addition, attorneys certified in elder law must be capable of recognizing issues of concern that arise during counseling and representation of older persons, or their representatives, with respect to abuse, neglect, or exploitation of the older person, insurance, housing, long term care, employment, and retirement. The certified elder law attorney must also be familiar with professional and non-legal resources and services publicly and privately available to meet the needs of the older persons, and be capable of recognizing the professional conduct and ethical issues that arise during representation. http://www.nelf.org/randregs.htm#howis (visited February 5, 2008).

[5] It must be recognized that there are various reasons why an attorney chooses to handle certain types of cases, and it truly is an over-simplification to give one reason.

Elder Law as a practice constantly changes and grows to meet client needs and respond to the evolution of programs, services and laws As Professor Frolik describes it in Chap. 2: "the reality is that the practice of elder law is a rich mosaic of legal planning that is continually evolving to better meet client legal, financial and social needs and concerns" (Frolik 2008 at 11).

Dr. Doron has written in Chap. 5 (and previously) about the five dimensions of a legal practice for elders (Doron 2008). A combination of multi-dimensional, therapeutic approaches has some benefit when applied to the issues faced by clients in an Elder Law practice. An Elder Law practice is well-served by a multi-dimensional legal approach, but perhaps should not be limited to five legal dimensions. There is a need to study and understand how other disciplines can play a role in an Elder Law practice.

Many attorneys in the United States have already incorporated a multi-disciplinary approach to client problems by using other professionals to help clients navigate the terrain of services and programs. Some United States Elder Law attorneys view multi-disciplinary practices favorably,[6] and may have care managers or nurses as members of their staff. Some Elder Law practices in the United States are evolving into full-service firms that provide legal work, "financial planning and "life-care" planning [for the] client [to] meet the financial and care needs..." (Frolik Chap. 2 at 11). Some firms offer clients "lifetime provision of long-term care advice and supervision" (Frolik Chap. 2 at 29). As described in Chap. 1, the Bogutz and Gordon firm is an example of a multi-disciplinary firm:

As the practice has grown, serving as fiduciary for these clients has expanded to the point at which my firm is now multidisciplinary, with lawyers, care managers who are social workers or registered nurses, financial managers who assist with financial management, bill paying, tax compliance and providing investment oversight for the clients' resources. In addition to serving as personal guardian or estate manager for clients with no one able to serve, we have also been assisting in families where there are conflicts as to the type of care to be provided and have been able to act as a disinterested professional in families with dysfunctions (Bogutz Chap. 1 at 3).

10.1 Education and Training

Proficient Elder Law attorneys will have acquired knowledge of some of the physiological and psycho-social challenges faced by many of their clients. In fact, as noted by NAELA, an Elder Law attorney understands how these challenges affect a client on a day to day basis in order to filter out stereotypes in learning more about the client's legal problem. Once the context is understood, the Elder Law attorney may

---

[6] See, *e.g.*, NAELA Aspirational Standards § D *Competent Legal Representation*, http://www.naela.org/pdffiles/AspirationalStandards.pdf.

then develop solutions to legal problems that also accommodate or work to eliminate these physical and psycho-social problems.[7]

10.2 Multi-Dimensional Approach and Solution

Elder law is multi-faceted and, as Dr. Doron suggests, needs a multi-dimensional model to respond to the client's needs:

Elder law, so the multi-dimensional model suggests, includes distinct types of legal tools, a range of political and philosophic approaches, and multiple perspectives on the concept of "elder rights." In other words, this field is by no means directed by a single viewpoint: it cannot be considered either "individualistic" or "paternalistic," nor can it be portrayed as promoting "negative" as opposed to "positive" rights, or as favoring the individual over the family, the private over the public. Elder law covers the range of possible approaches and perspectives, and it can be fully appreciated only by understanding the internal balance between all of its components (Doron Chap. 5 at 70).

An adversarial model of jurisprudence, such as that in the United States, assumes opponents and conflict, rather than collaboration aimed at reaching a mutually desirable resolution. Applying this to Elder Law, autonomy or individualism may be opposite of paternalism or loss of rights; individuals may be on the opposite side of the application of social policy; and the elder may be at odds with the family (Doron Chap. 5). This is just one reason why Elder Law should be approached from a multi-disciplinary approach. If an attorney has a working understanding of how physical and psycho-social problems affect a client, the attorney may be able to craft a more effective solution for the client.

In order to implement a multi-disciplinary approach to problems that present as legal issues, consider how allied professionals are educated and trained. There may be little interaction or opportunity to work with professionals from other disciplines, although it is clear that at some point that clients will need solutions from various experts. Consider the ability to live independently as an example. A good house

[7] *See, e.g.*, NAELA Aspirational Standards § D *Competent Legal Representation*, http://www.naela.org/pdffiles/AspirationalStandards.pdf.

The NAELA website describes Elder Law practice this way:

Attorneys who primarily work with the elderly bring more to their practice than an expertise in the appropriate area of law. They bring to their practice a knowledge of the elderly that allows them and their staff to ignore the myths relating to aging and the competence of the elderly. At the same time, they will take into account and empathize with some of the true physical and mental difficulties that often accompany the aging process. Their understanding of the afflictions of the aged allows them to determine more easily the difference between the physical versus the mental disability of a client. They are more aware of real life problems, health and otherwise, that tend to crop up as persons age. They are tied into a formal or informal system of social workers, psychologists and other elder care professionals who may be of assistance.... http://www.naela.org/qa.aspx (visited February 7, 2008).

design can enhance independence for a frail elder. Geography is also important. A house might be elder-friendly, but if services and goods are inaccessible to the elder with limited transportation options, the elder may not be able to remain in the home. Yet how often do elder law attorneys, architects and urban planners meet to talk about the need for clients to age in place?

The policies that evolve from laws in the payment of and provision for long term care can impact Elder Law clients. Decisions regarding eligibility for, type and amount of available services are often made for budgetary reasons rather than providing funding and coverage for preventative care or supportive services. The United States health care system places less emphasis on providing preventative services than on treating a problem after it has occurred. Good public policy looks more broadly than dollars and cents in developing ways to meet the needs of those unable to do so.

10.3 Does the Current System Work?

Is there a need to rethink the way the legal system responds to the needs of elders (Doron, Chap. 5)? Would a better system be to use a "continuum of aging" rather than use a bright line age to define an Elder Law practice (*see e.g.,* Surtees Chap. 7)? Two examples of changing the way the legal system responds come to mind. First, some experts in the field, for example, advocate for the use of mediation in an elder law practice for dispute resolution (*see e.g.*, Karp and Wood 2000; Radford 2002; Larsen and Thorpe 2006). Secondly, because a client's personal situation (health, income, physical and mental abilities) affects the client's ability to function, in the United States, "modern" guardianship theory advocates the use of the least restrictive alternative[8] and limited guardianship.[9] As noted by Winsor Schmidt in Chap. 9, laws that deal with incapacity and mental health may reinforce stereotypes:

[t]he significance of mental health law concepts like legal capacity, mental competence, guardianship, and substitute decision making as a core around which elder law is established becomes more evident. These mental health law concepts, when used as instruments of sanism, if not of the therapeutic state, are reinforcing, congruent with, and complementary to, their use as tools of ageism, if not of the therapeutic state (Schmidt, Chap. 9 at 132).

Perhaps the next step is an examination of the role of law in Elder Law. Does the law apply to every situation? A dysfunctional family is a dysfunctional family. The law may not remedy that, but the law can redress wrongs inflicted by one family member upon another. Is there a lack of personal responsibility in families to provide care for their elder relatives? Should countries adopt the model of supporting

---

[8] *See, e.g.*, Uniform Guardianship & Protective Proceedings Act, §§ 304, 305.

[9] *See, e.g.*, Uniform Guardianship & Protective Proceedings Act, §§ 304, 305. See also Kapp (2008, pp. 7–9).

caregivers in informal support networks in order to provide better care to the elders (Doron, Chap. 5)?

It is commonly observed that society can not effectively legislate morality. Can we mandate that families provide proper care for their elders?[10] If it is true that we all have a moral obligation to provide care for our own elder family members, does the law act responsibly if it provides financial incentives for that care? Would it be sound social policy (Doron, Chap. 5)? Legal systems, lawmakers, policymakers and leaders must reexamine the law's responses to and solutions for the problems faced by the world's elders. Gender- and age-based attitudes are reflected in laws and policies; in order to prepare to meet the needs of this growing segment of the world's population, those biases need to be addressed (Dayton 2008, pp. 75–76). Professor Surtees argues that rather than using age as a dividing line in determining whether a person is an elder, we begin to think in terms of a "continuum of age" by using universalism to understand elder law, and to "[design] programs and policy which include us all, wherever we currently find ourselves on time's continuum" (Surtess, Chap. 7 at 105).

10.4 As to the Future: Where Do We Go From Here?[11]

The practice of Elder Law has experienced some robust growth over the past twenty years. As noted earlier, Elder Law has become a general practice area within which one specializes. Second-generation clients will soon be a commonplace, as the children of elder clients turn to their parents' Elder Law attorneys for representation with their legal problems.

Will the practice of Elder Law continue its boom? Elder Law is a growing field in some countries, while in other countries elders may wrestle with more basic problems, such as food, clothing, shelter and health care. Life expectancy may not be uniform throughout the world but as life expectancy increases, the issues of

[10] For a discussion about filial or family responsibility, *see* Pakula (2005). There are those in the United States who would be opposed to laws that impose a personal financial duty to support one's parents. In reality many already provide financial or other support to their parents.

Although it may be possible to pass laws that require families to provide financial support for their elders' care, there may still be issues regarding proper care, and it is still impossible to legislate that families love and care. Elder Abuse is a growing and serious problem in the United States and current research shows that oftentimes the abuser is a family member-caregiver. According to the National Center on Elder Abuse, "[f]amily members are more often the abusers than any other group. For several years, data showed that adult children were the most common abusers of family members; recent information indicates spouses are the most common perpetrators when state data concerning elders and vulnerable adults is combined." http://www.ncea.aoa.gov/NCEAroot/Main_Site/FAQ/Questions.aspx

[11] For a detailed discussion of the future of Elder Law in the United States, see Morgan (2007)

aging are likely to be more universal. That worldwide growth in the population of elders will increase the need for Elder Law attorneys.

As individuals live to advanced ages, they may face a number of legal problems (Frolik, Chap. 2). As a result, those of advanced age will be, in Professor Frolik's words, the "foci" of an Elder Law practice (Frolik, Chap. 2 at 11). As people live longer, they may experience more health problems, which in turn may well increase their need for legal assistance. Those individuals need help with planning for dementia[12] as well as planning for the real possibility of a much longer *and* healthy and active life.

It would also be well to remember human nature. It is not unusual for people to fail to plan, and to seek legal help only after a crisis. Hope springs eternal, as does procrastination. Dysfunctional families, furthermore, do not often become happy and close-knit as a result of the aging of a parent. More and more nations, societies and communities operate by rule of law, and a handshake may not be enough to make a workable and binding agreement.

Reliance on government programs will be replaced by an increasing emphasis on personal responsibility and "pay as you go" at least for social service programs and long-term care in the United States. Changes in United States programs and services are often driven by a policy of budget-cutting. As noted in Chap. 9 by Prof. Schmidt:

One critical view of law and aging, or elder law, sees a dramatic historical shift of public policy about the elderly from concerns about equity and social justice to efficiency and cost containment. While Social Security, Medicare, and Medicaid achieved a sea change in the incidence of poverty and quality of life for the elderly, the preoccupations with cost containment in the late twentieth century arguably changed the function and political economy of elder law and policy from social justice to social control (Schmidt, Chap. 9 at 122).

As elders live longer, will social programs need to change eligibility and coverage in order to meet the needs of clients? If social programs make such fundamental changes, then clients will need Elder Law attorneys even more.

Life is complex. Often clients must make significant decisions concerning the remaining years of their lives: where to live, how to pay for health care, when to stop medical treatment, or who to designate as a surrogate decision-maker. Regardless of the theory or approach, clients need Elder Law attorneys to navigate the issues as they live the final phase of their lives. Do policy makers need to re-think the approach to aging or continue on this current collective course? The demographic shifts with the aging of the Baby Boomers[13] make it unlikely, at least in the United States, that services and programs can be funded and provided at current levels.

[12] See Frolik, Chap. 2. Professor Frolik refers to "dementia planning" as a "sub-speciality" of an Elder Law practice.

[13] In the United States the Baby Boomers, in discussing the demographics, are often referred to as the pig in the python or the silver tsunami Professor Dayton in her chapter makes reference to "apocalyptic demographics" Dayton, Chap. 4.

Professor Dayton observes that "[m]ost developed nations have been compelled by the convergence of these demographic realities to look for new ways to deliver, and pay for, the expanded need for long term care" (Dayton, Chap. 4 at 52).

Clients see Elder Law attorneys when they have problems and need solutions, or to plan for the future. An Elder Law practice can be as broad or as focused the attorney chooses. With the breadth of Elder Law practice, it is important to develop a necessary expertise in specific areas of Elder Law to serve clients.

Rather than a compartmentalized approach to aging, a better approach might be to examine the continuum of issues and services, as well as the way we age. A preventative, planning, therapeutic approach should help avoid, or at least minimize the impact of, legal matters that clients might otherwise endure.

Elder Law practice can benefit from an examination of the various theoretical approaches to law and aging. For example, Professor Kaplan offers one advantage to the application of law and economics theory to elder law ("assessing whether certain statutes accomplish their intended objectives or simply confound their presumptive beneficiaries and frustrate sensible public policy") (Kaplan, Chap. 6 at 91) while also noting that in some instance the application of a cost-benefit analysis would simply be implausible (Kaplan, Chap. 6 at 76 (quoting Thomas Ulen)). One aspect from a disability law perspective is incorporation of the examination of how the law interacts with "member of the… community", whether positively or negatively (Surtees, Chap. 7 at 94).

Professor Dayton supports the concept of selecting the best from these various approaches:

No single theoretical approach to elder law is capable of producing complete justice for all persons affected by law and aging policy. Aging affects individuals differently depending on many factors–economic status, gender, race, disability, and so on. It is important, therefore, that all perspectives… be included in policy discussions about law and aging and in legal texts that implement aging policy. Only then will "elder law" embody a rational and just account of the relationship between aging of the body and the positive law (Dayton Chap. 4 at 56–57).

Alan Bogutz, in Chap. 1, offers his view of the future of an elder law practice:

[I] think the future of the elder law practice requires lawyers to expand their skills to the other substantive issues of interest to older persons. This, I believe, is particularly important as the next generation of senior Americans, The Boomers, reaches age 60 and beyond … This is a different generation from those who lived through the Depression of the 1930's, a different generation from those lived and fought in World War II and a different generation from the cohort that came of age in the 1950's. The Boomers are the generation that matured in the turbulent 1950's and 1970's and share very different perspectives on life, retirement, saving and the entitlement to governmental programs. This is likely to be a much more demanding, much more litigious and much more activist group of older persons than their predecessors. At the same time, this generation is no longer just the "Sandwich Generation" caring for themselves and their children as well as their parents – they are becoming the "Club Sandwich Generation," caring also for their own grandchildren in many instances. This is the next great wave of our clients…

…[T]here will be far greater need for elder law attorneys to serve this generation and to be their advisors and advocates in many more areas than most elder law attorneys are now practicing. The future of elder law should include much more emphasis on issues of age discrimination in employment, in credit and in housing. Future elder law attorneys are also more frequently going to be looked to for help with managing finances and personal care, providing a more complete range of services for clients who have neither family nor friends to help with their daily needs. Recovery of assets for those who have been exploited will become more of our practice. Finally, the elder law attorneys should become more familiar with the laws concerning the abuse and neglect of elders in their own homes or in institutional settings – this becomes more important as greater numbers of elders are cared for by strangers. We need to be prepared to write and review care contracts, examine and recommend changes on institutional care agreements and be creative in developing options in a time when there will be more persons in need of care than there will be available caregivers (Bogutz, Chap. 1 at 8).

Age does not necessarily change what a person believes, but it may change the person's abilities. Much can be learned from the evolution of disability law in designing programs and services to meet the needs of today's – and tomorrow's – elders. Everyone benefits from universal design[14] and from accommodations made to give access to those individuals with differing abilities. Aging is a great universal – everyone does it – every day, but everyone does it differently. No one solution, no one theory, no one-size-fits-all approach to Elder Law will serve. Instead, take the best of all and craft the best solution – or solutions – for the client – or clients. Remember that aging is universal – since everyone does it every day, each person has a vested interest in the future of Elder Law.

References

Karp, N, Wood E (2000) Building coalitions in aging, disability and dispute resolution. ABA

Larsen R, Thorpe C (2006) Elder mediation: optimizing major family transitions. Marquette's Elder Advisor 7:293

Morgan, RC (2007) Elder law in the United States: the intersection of the practice and demographics. J Int Aging, L Pol 2:103. http://www.law.stetson.edu/centers/AARPJournal/

Needham, HC Elder law comes of age. http://www.nelf.org/elderlaw.htm. Accessed on 5 February 2008

Pakula, M (2005) A federal filial responsibility statute: a uniform tool to help combat the wave of indigent elderly. Family L Q 39:859

Radford, MF (2002) Is the use of mediation appropriate in adult guardianship cases? Stetson, L Rev 31:611

[14] *See e.g.*, the Center for Universal Design, http://www.design.ncsu.edu/cud/ (visited February 8, 2008).

Index

Printing: Krips bv, Meppel, The Netherlands
Binding: Stürtz, Würzburg, Germany